# Applications of Artificial Intelligence in Additive Manufacturing

Sachin Salunkhe
*Vel Tech Rangarajan Dr.Sagunthala R&D Institute of Science and Technology, India*

Hussein Mohammed Abdel Moneam Hussein
*Helwan University, Egypt*

J. Paulo Davim
*University of Aveiro, Portugal*

A volume in the Advances in Computational Intelligence and Robotics (ACIR) Book Series

IGI Global
PUBLISHER of TIMELY KNOWLEDGE

Published in the United States of America by
IGI Global
Engineering Science Reference (an imprint of IGI Global)
701 E. Chocolate Avenue
Hershey PA, USA 17033
Tel: 717-533-8845
Fax: 717-533-8661
E-mail: cust@igi-global.com
Web site: http://www.igi-global.com



Library of Congress Cataloging-in-Publication Data

Names: Salunkhe, Sachin, 1984- editor. | Hussein, Hussein Mohammed Abdel Moneam, 1966- editor. | Davim, J. Paulo, editor.
Title: Applications of artificial intelligence in additive manufacturing / Sachin Salunkhe, Hussein Mohammed Abdel Moneam Hussein, and J. Paulo Davim, editor.
Description: Hershey, PA : Engineering Science Reference, an imprint of IGI Global, [2022] | Includes bibliographical references and index. | Summary: "This book provides introductory instruction on how to learn how to use artificial intelligence to produce additively manufactured parts, including a description of the starting points, what you can know, how it blends and how artificial intelligence in additive manufacturing apply"-- Provided by publisher.
Identifiers: LCCN 2021036269 (print) | LCCN 2021036270 (ebook) | ISBN 9781799885160 (h/c) | ISBN 9781799885177 (s/c) | ISBN 9781799885184 (ebook)
Subjects: LCSH: Additive manufacturing--Data processing. | Artificial intelligence--Industrial applications.
Classification: LCC TS183.25 .A66 2022 (print) | LCC TS183.25 (ebook) | DDC 621.9/88--dc23/eng/20211012
LC record available at https://lccn.loc.gov/2021036269
LC ebook record available at https://lccn.loc.gov/2021036270

This book is published in the IGI Global book series Advances in Computational Intelligence and Robotics (ACIR) (ISSN: 2327-0411; eISSN: 2327-042X)

British Cataloguing in Publication Data
A Cataloguing in Publication record for this book is available from the British Library.



For electronic access to this publication, please contact: eresources@igi-global.com.

# Advances in Computational Intelligence and Robotics (ACIR) Book Series

ISSN:2327-0411
EISSN:2327-042X

Editor-in-Chief: Ivan Giannoccaro University of Salento, Italy

**MISSION**

While intelligence is traditionally a term applied to humans and human cognition, technology has progressed in such a way to allow for the development of intelligent systems able to simulate many human traits. With this new era of simulated and artificial intelligence, much research is needed in order to continue to advance the field and also to evaluate the ethical and societal concerns of the existence of artificial life and machine learning.

The **Advances in Computational Intelligence and Robotics (ACIR) Book Series** encourages scholarly discourse on all topics pertaining to evolutionary computing, artificial life, computational intelligence, machine learning, and robotics. ACIR presents the latest research being conducted on diverse topics in intelligence technologies with the goal of advancing knowledge and applications in this rapidly evolving field.

**COVERAGE**

- Natural Language Processing
- Intelligent Control
- Fuzzy Systems
- Cognitive Informatics
- Artificial Life
- Brain Simulation
- Algorithmic Learning
- Automated Reasoning
- Computational Logic
- Neural Networks

IGI Global is currently accepting manuscripts for publication within this series. To submit a proposal for a volume in this series, please contact our Acquisition Editors at Acquisitions@igi-global.com or visit: http://www.igi-global.com/publish/.

The Advances in Computational Intelligence and Robotics (ACIR) Book Series (ISSN 2327-0411) is published by IGI Global, 701 E. Chocolate Avenue, Hershey, PA 17033-1240, USA, www.igi-global.com. This series is composed of titles available for purchase individually; each title is edited to be contextually exclusive from any other title within the series. For pricing and ordering information please visit http://www.igi-global.com/book-series/advances-computational-intelligence-robotics/73674. Postmaster: Send all address changes to above address. 

## Titles in this Series

*For a list of additional titles in this series, please visit:*
*www.igi-global.com/book-series/advances-computational-intelligence-robotics/73674*

***Socrates Digital™ for Learning and Problem Solving***
Mark Salisbury (University of St. Thomas, USA)
Engineering Science Reference • © 2022 • 383pp • H/C (ISBN: 9781799879558) • US $215.00

***Regulatory Aspects of Artificial Intelligence on Blockchain***
Pardis Moslemzadeh Tehrani (University of Malaya, Malaysia)
Engineering Science Reference • © 2022 • 273pp • H/C (ISBN: 9781799879275) • US $245.00

***Genetic Algorithms and Applications for Stock Trading Optimization***
Vivek Kapoor (Devi Ahilya University, Indore, India) and Shubhamoy Dey (Indian Institute of Management, Indore, India)
Engineering Science Reference • © 2021 • 262pp • H/C (ISBN: 9781799841050) • US $225.00

***Handbook of Research on Innovations and Applications of AI, IoT, and Cognitive Technologies***
Jingyuan Zhao (University of Toronto, Canada) and V. Vinoth Kumar (MVJ College of Engineering, India)
Engineering Science Reference • © 2021 • 570pp • H/C (ISBN: 9781799868705) • US $325.00

***Decision Support Systems and Industrial IoT in Smart Grid, Factories, and Cities***
Ismail Butun (Chalmers University of Technology, Sweden & Konya Food and Agriculture University, Turkey & Royal University of Technology, Sweden)
Engineering Science Reference • © 2021 • 285pp • H/C (ISBN: 9781799874683) • US $245.00

*For an entire list of titles in this series, please visit:*
*www.igi-global.com/book-series/advances-computational-intelligence-robotics/73674*

IGI Global
PUBLISHER of TIMELY KNOWLEDGE

701 East Chocolate Avenue, Hershey, PA 17033, USA
Tel: 717-533-8845 x100 • Fax: 717-533-8661
E-Mail: cust@igi-global.com • www.igi-global.com

# Table of Contents

# Detailed Table of Contents

approach to fabricate the test specimens. The optimal SIS parameters are obtained through a swarm intelligence metaheuristic technique namely dragonfly algorithm (DFA). The optimal parameter settings of LT of 0.102 mm, HE of 28.46 J/mm2, HFR of 3.22 mm/sec, and PFR of 110.49 mm/min are achieved through DFA for improved tensile and flexural strengths of 26.21 MPa and 65.71 MPa, respectively. Further, the prediction ability of DFA was compared with particle swarm optimization algorithm.


*Aya Abd Alla Ramadan, Faculty of Engineering, Helwan University, Egypt*
*Sherif Elatriby, Faculty of Engineering, Helwan University, Egypt*
*Abd El Ghany, Faculty of Engineering, Helwan University, Egypt*
*Azza Fathalla Barakat, Faculty of Engineering, Helwan University, Egypt*

This chapter summarizes a PhD thesis introducing a methodology for optimizing robotic MIG (metal inert gas) to perform WAAM (wire and arc additive manufacturing) without using machines equipped with CMT (cold metal transfer) technology. It tries to find the optimal MIG parameters to make WAAM using a welding robot feasible production technique capable of making functional products with proper mechanical properties. Some experiments were performed first to collect data. Then NN (neural network) models were created to simulate the MIG process. Then different optimization techniques were used to find the optimal parameters to be used for deposition. These results were practically tested, and the best one was selected to be used in the third stage. In the third stage, a block of metal was deposited. Then samples were cut from deposited blocks in two directions and tested for tension stress. These samples were successful. They showed behavior close to base alloy.


*Prafull Agarwal, CAD Lab, Department of Mechanical Engineering, Manipal University Jaipur, India*
*Rishi Kurian, CAD Lab, Department of Mechanical Engineering, Manipal University Jaipur, India*
*Ravi Kumar Gupta, Manipal Institute of Technology Bengaluru, Manipal Academy of Higher Education, Manipal, India*

Additive Manufacturing (AM) is a layer-by-layer deposition of material for the production of the desired product. The design flexibility associated with AM is much more when compared to the conventional manufacturing process. To manufacture a part with AM, two things play a critical role: the designing of the part and the other is the placement of the part in the build volume. As already mentioned, design flexibility associated with AM is much more when compared to the conventional manufacturing process. However, to correctly implement the design flexibility, we need a knowledge base at our disposal so that appropriate features can be used for the part production. The AM feature taxonomy forms the backbone of the knowledge base. The taxonomy comprises AM features classified based on different categories, which helps us understand every feature's importance. Talking about the part placement, we know that optimal placement is the key factor that makes the AM process economically feasible.


*Sachin Salunkhe, Vel Tech Rangarajan Dr. Sagunthala R&D Institute of Science and Technology, India*
*Vishal Gangadhar Naranje, Amity University, Dubai, UAE*
*Jayavelu S., Vel Tech Rangarajan Dr. Sagunthala R&D Institute of Science and Technology, India*
*Atiq Rehman, Amity University, Dubai, UAE*

Over the past decades, air traffic has increased to such an extent that it has highly impacted (anthropogenic) climate change due to heat, noise, particulates, and gas emissions. With airplane turbines being a pivotal contributor to such adverse developments, there has been an increasing interest in research regarding the optimization of airplane turbines. In line with these efforts, this chapter adopts an innovative approach that bridges the digital and physical through the application of digital twin technology to direct metal laser melting to optimize product development. Specifically, it encompasses a guideline towards how digital twin solutions are created based on all the latest research. A manual approach devises a digital twin interface where the prototype is manufactured used additive manufacturing. This manual can then be applied to optimize airplane turbines regarding their safety, environmental impact, fuel efficiency, and cost.

# Preface

Computer-connected devices that mimic human intelligence are referred to as artificial intelligence (AI). Additive manufacturing (AM) applications can currently be found in the food, chemical, aerospace, automotive, and healthcare industries, among other sectors. Perhaps the most significant advantage of 3D printing is that the customer's specifications can create even complex objects. In the current state of affairs, it is better suited for small-scale production. 3D model preparation, component prototyping, and component production are the stages of additive manufacturing. In the prefabrication stage, the goal is to determine whether or not it is technically possible and feasible to print a given 3D model. It is also known as smart manufacturing when 3D printing with artificial intelligence is used. Productivity would increase as a result of smart manufacturing.

It is predicted that the global 3D printing market will grow to $6 billion by 2022, with the biggest growth opportunities for businesses in the home improvement and life sciences industries. Although the additive manufacturing process has made significant strides in recent years, there are still several obstacles to overcome before being widely adopted by the industry. Taking additive manufacturing as an example, there are numerous and complex variables that must be monitored and controlled throughout the process to achieve a reasonable level of print accuracy. Experimenting with different lattice positions or designing appropriate support structures is not a sustainable or time-efficient method of finding the best configuration. Machine learning is currently being used to solve this problem through generative design and testing in the pre-fabrication stage, to increase printing efficiency and cost savings while also improving print quality. Intelligent service-oriented production processes for the industrial sector are being developed using artificial intelligence in 3D printing and additive manufacturing at this time.

In this book, we look at how artificial intelligence is being used in the additive manufacturing and 3D printing industries today. During our research, we found that today and shortly (within the next two years), AI applications seem to cluster around the following broad categories: Increasing prefabrication efficiency, detecting defects, Monitoring construction in real-time, and Predictive maintenance.

Artificial intelligence applications in additive manufacturing are discussed in the book, with particular attention being paid to aerospace parts optimization, automotive, consumer products, industrial products, medical devices, and other fields. All of the chapters in this book present the results of the authors' years of hard work and dedication in conducting their research. It provides an overview of the most recent research developments in the field, but it also identifies the future scope of work for young graduates and engineers and senior R&D professionals working in the area of additive manufacturing. There is no doubt that the research work reported in the various chapters of this book will be beneficial to the additive manufacturing industries and other industries.

This book, *Applications of Artificial Intelligence in Additive Manufacturing*, not only highlights the most recent research status in the domain but also identifies potential areas of work for young researchers, mechanical and materials engineering students, and professionals who are interested in learning the fundamentals and most recent applications of artificial intelligence in additive manufacturing applications.

The book is divided into nine chapters, each of which was written by an expert. The following sections provide a brief overview of the various chapters contained within the book.

The introduction represents an introduction to artificial intelligence in additive manufacturing. The detailed discussion about different types of AI techniques used in additive manufacturing such as artificial neural network (ANN), machine learning (ML), adaptive neuro fuzzy inference system (ANFIS) and evolutionary algorithm (EA) are presented.

Chapter 1 presents an optimization of additive manufacturing process such as design, process plan and control, process monitoring, quality enhancement of additively manufactured products using machine learning algorithm.

Chapter 2 discussed a deep learning neural network (DLNN) to rationalize the effect of cellular structure design factors as well as process variables on physical and mechanical properties utilizing Laser Powder Bed Fusion. For both design and process optimization, the trained deep learning neural network model are also explained.

Chapter 3 focuses on prediction of surface roughness which effected by the building parameters such as layer thickness and building orientation using regression and artificial intelligent models. The outcome of developed model show that the layer thickness parameter has an effect more than the building orientation parameter on the surface roughness are presented.

Chapter 4 developed machine learning model to numerous aspects of the additive manufacturing whole chain, counting model design, and quality evaluation especially directed energy deposition (DED), powder bed fusion (PBF), and material extrusion process. Further, Present challenges in applying Machine learning to additive manufacturing and possible solutions for these problems are discussed.

Chapter 5 complies on utilizing a hybrid approach of response surface methodology and dragonfly algorithm for investigations and optimization of the selective inhibition sintering (SIS) process to improve the mechanical strengths such as tensile and flexural of fabricated high density polyethylene parts.

Chapter 6 describes the methods of optimization of robotic metal inert gas to perform wire and arc additive manufacturing using artificial neural network and genetic algorithm.

Chapter 7 Comprising taxonomy additive manufacturing features classified based on different categories such as deform, cut, transition, and compound are discussed. The proposed taxonomy useful for the classifications of feature based additive manufacturing parts.

Chapter 8 proposed model for process optimizations of direct metal laser melting using digital twin technique for the application of airplane turbines. The detailed discussion about innovative approach that bridges the digital and physical through the application of digital twin technology to Direct Metal Laser Melting to optimize product development are presented.

Chapter 9 presents online prediction and detections of fused deposition modeling parts using machine learning and convolutional neural network model. The proposed model, which is based on a machine learning model, detects defects such as delamination online monitoring system.

Additive manufacturing stands to benefit greatly from artificial intelligence (AI) implementations. AI applications in streamlining additive manufacturing to be integrated into other manufacturing techniques or become a commodity for users are still far in the future, despite some progress having been made in this area. Design correlation, design improvement, defect reduction, and microstructural design are areas in which AI can help additive manufacturing. However, the main hurdle at the moment is the availability and reliability of data that are required for training. AM can significantly benefit from current approaches developed for other processes; the available experimental data has a wide range and is not always open to the public, thanks to the research community or AM manufacturers. In the development of AM-specific AI algorithms, ensuring the integrity of the data is critical. Experiments have been conducted or are currently being conducted with wide variations in observation and results. Creating a data storage facility is therefore critical for the manufacturing industry. The data generation condition must be disclosed along with the data to work correctly for the AI algorithms. Information such as process parameters, exact details of raw materials such as powder or feedstock, composition, raw material properties such as flowability of powder, particle size, etc., and the type of machine should be disclosed and shared to recreate label the data. In addition, the process's difficulties, such as high temperatures or high speeds, make it challenging to monitor and measure. The majority of currently available technologies rely on imaging the material's surface thermally or optically, and depth information is rarely readily available. Companies and academics must work together to create tools that accurately and quickly track a wide range of process conditions and parameters. For example, it's not yet possible to detect defects, create a 3D image of the build while it's being built, or keep track of the microstructure and grain orientations. These areas would undoubtedly present some difficulties that a willing and motivated community can overcome.

# Introduction

Three-dimensional (3D) printing or additive manufacturing (AM) techniques have matured and ushered in a new product design and production era. Parts with complex geometries and functionally graded properties can be made using layer-by-layer fabrication techniques. AM also saves resources because they use less material. End-part production is slowly but surely replacing prototyping as the primary use of the technology. In order to print real functional parts with various types and forms of materials, various AM fabrication techniques have been developed, such as fused filament fabrication, stereolithography, selective laser sintering, selective laser melting, and laser engineered net shaping. To be sure, there are several unique difficulties to be overcome, such as the materials' anisotropy and porosity from poor fusion and warping from residual stress caused by rapid cooling in additive manufacturing (AM). Comprehensive knowledge of the AM process is required, from the processability of the feedstock materials (such as rheological properties and powder flowability) to the relationship between the AM part's process-structure-property attributes. Several process parameters in AM can impact the quality of the final parts, necessitating a multidisciplinary approach that considers material properties, solid-liquid interactions, fluid dynamics, grain growth, and thermal-mechanical interactions. In order to create physics-based models, one must have a thorough understanding of additive manufacturing processes on multiple scales and in multiple physics. Therefore, individual research usually only covers a few aspects of the printing process as a whole and thus restricts its representation. Microscale grain structure evolution of powder bed fusion using computational fluid dynamics (CFD), analytical modeling of residual stress, and macro-scale melt pool profile and bead shape using finite element analysis have been attempted. Such investigations take much time. Using physics-based numerical simulations to emulate the entire additive manufacturing process is therefore tricky. Because AM process optimization can be performed with only partial or incomplete information, machine learning (ML) data-driven models based on physical understanding of AM processes are critical. Parts made for practical use must meet specific quality and reliability requirements, and quality control of AM parts has received significant attention from the industries they are intended for. With the development and use of more advanced AM materials in structurally critical part's high part quality must be ensured. Unwanted porosity in AM processes is well-documented. Proses have a significant impact on the mechanical properties of additively manufactured (AM) parts. According to the results of one study, a well-controlled system can produce an extremely dense AM part (density >99.8%). To improve AM parts' quality, in-situ quality control techniques must be used to detect anomalies and print corrections using closed-loop feedback. In the AM field, data-driven models have overcome time-consuming physics-based modeling and detect anomalies during in-process monitoring for quality control. In order to predict certain behaviors and properties, the ML algorithms collect and process a large amount of data. When used in additive manufacturing (AM), it can identify patterns or irregularities in the manufacturing process to be corrected. All aspects of AM, the fabrication process and qualification, have been significantly influenced by ML. The impact of machine learning (ML) is only going to increase in the future. With today's large amount of data, abundant computational resources, and advanced algorithm structure, the neural network (NN) algorithm is the most widely used and under rapid development among ML methods (LeCun et al., 2015). When it comes to computer vision (Krizhevsky et al., 2012), voice recognition (Anusuya & Katti, 2010), natural language processing (Devlin et al., 2018), and autonomous driving, for example, NNs are the driving force. Most of these tasks were once thought to be only possible by humans until the NN showed its great power in recognizing the complicated underlying patterns. There is also a trend in the use of NN in traditional manufacturing fields due to the positive results. Product design, manufacturing, qualification, and delivery have all been impacted by NN, and it is expected that this impact will grow in intensity over time.

## METHODS

Three-dimensional printing (3DP) is now widely used in a wide range of fields such as medical technology, architecture, food technology, aeronautics, chemical manufacturing, education, and society. However, there are still limitations in today's manufacturing industry that prevent 3DP from becoming more widely used. Computational intelligence, or so-called artificial intelligence, is needed to solve the problems of excessive time consumption, a lack of real-time control, potential security risks, and the shift from mass production to mass customization (AI). Many fields, such as aviation, computer science, finance, education, healthcare, and medicine, have all used AI successfully. Because of the ever-increasing data repository, it provides a wide range of algorithmic, theoretical, and methodological options that have the potential to transform current manufacturing standards for the better. For example, artificial neural networks, machine learning, adaptive neuro-fuzzy inference systems, evolutionary algorithms, etc.

## Artificial Neural Network Algorithm

The 3D printing process includes model design, material selection, printing, and part evaluation and characterization. This section combines the use of ANN for 3D printing process monitoring, design, and correlation between process parameters and final component characteristics.

### Process Surveillance

Process monitoring via various sensors provides direct information regarding quality supervision and control during the printing process. Three distinct types of data sources can be distinguished: (a) one-dimensional data sources such as spectra; (b) two-dimensional data sources such as graphs and images; and (c) three-dimensional data sources such as morphologies (Everton et al., 2016). While one-dimensional data is faster to process, it contains less information than two- and three-dimensional data. Shevchik et al. (Shevchik et al., 2018; Wasmer et al., 2019) investigated the possibility of using acoustic emission to monitor product quality by using an acoustic emission sensor in conjunction with an ANN.

### Designing

Design is a critical area of research that necessitates a thorough understanding of the capabilities and limitations of 3D printing technologies. It is the initial and most critical step in the workflow process. A well-designed CAD model not only ensures printability but also minimizes the amount of support material required. The design process, on the other hand, is typically iterative and time-consuming. A data-driven approach to 3D printing design would aid designers in their work. Maidin et al. (Bin Maidin et al., 2012) demonstrated that the design feature database provided novice designers with ideas and design features. Applying the ANN technique in 3D printing enables feature recommendations for existing CAD models, assisting designers in expediting the design stage decision-making process. For example, Yao et al. (Yao et al., 2017) developed a hybrid ANN algorithm that utilized hierarchical clustering to classify 3D printing design features and a support vector machine (SVM) to enhance the hierarchical clustering results in pursuit of recommended design features. It aided inexperienced designers unfamiliar with 3D printing in determining appropriate design features for remote-controlled car components without the need for physical trials and errors. Apart from that, ANN algorithms have been used to recognize features in CAD models and to analyze the manufacturability of 3D printed objects. Shi et al. used the heat kernel signature and multiscale clustering to identify potential design flaws in a particular CAD model (Shi et al., 2018). Zhang et al. used a double-layered extreme ANN (DLEANN) to determine the optimal print orientation to avoid placing support structures on user-preferred features (Zhang et al., 2015).

The first layer in this DLEANN was classification, which was used to determine the relative score between the various part orientations, and the second layer was a regression used to create a global score for all printing directions. It was discovered to identify the optimal printing directions with the fewest visual artifacts associated with support removal. Pre-processing data is critical for data-driven ANN because it removes incorrect data and trains it with the correct data. The data must be filtered so that an ANN model only uses cracks as input from scanning electron images of deposited layers. However, extracting the precise locations of the cracks distributed along the grain's periphery. Extracting precise information is a difficult task for those without reliable data and image processing experience. Data preprocessing standards and practices must be developed as a result.

## Machine Learning (ML)

As shown in Figure 1, it is beneficial to break down the applications into the pre-processing, process, and post-process for the application of AI in AM, given the complexity of the process. Design space can use ML during the pre-processing stage (geometrical design, topology optimization, raw materials design, and powder properties). Predicting material properties is now possible thanks to advances in machine learning (ML) in raw material design. Designing new novel materials is also made easier with it (Gu et al., 2018; Li et al., 2016), and it can take advantage of AM's unique manufacturing capabilities to bring designs to life that otherwise would not be possible. Even though machine learning has been applied to design space (geometric design and topology optimization), it has not been applied to powder properties, which are the least explored areas. The process itself categorizes ML applications for experimentation in-process monitoring and optimization, and simulation work in the same area. One of the areas that have seen much research is experimental process monitoring and optimization.

## Adaptive Neuro-Fuzzy Inference System (ANFIS)

One of the soft computing techniques, adaptive neuro-fuzzy inference system (ANFIS), plays a significant role in accurately modeling input-output matrix relationships (Savkovic et al., 2019). Because of the nonlinearity of the 3D printed

*Figure 1. AM powder bed manufacturing process (Ladani, 2021)*

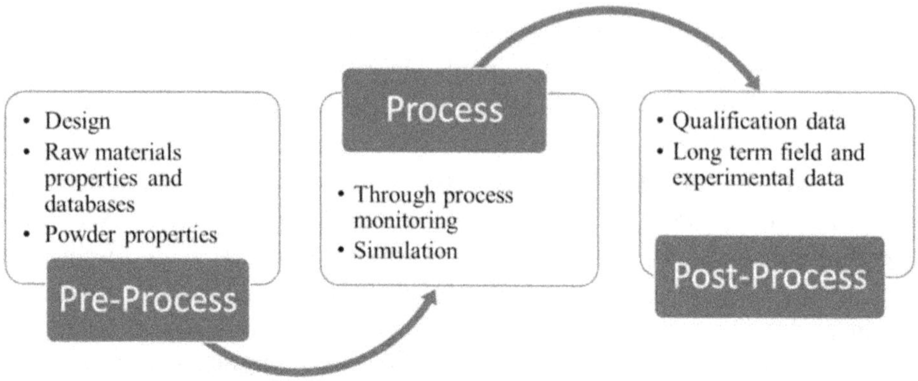


PLA process, ANFIS is an excellent tool for predicting weight gain based on input variables. ANFIS has also been utilized in the modeling and evaluation of metallic slurry erosion. According to Hassan et al. (Hassan et al., 2012), a fuzzy model was developed to evaluate and predict 5127 steel slurry erosion. The model they created matched experimental results quite well. Ramesh et al. (Ramesh et al., 2013) used a fuzzy logic approach to predict the slurry erosive wear behaviour of Al6061 alloy and found that the predicted values were very close to the experimental results. Shamshirband et al. (Band et al., 2015) developed an ANFIS model to estimate the erosion rate of copper particles flowing through an aluminum 3003 alloy elbow. It was found that the ANFIS model had a high degree of accuracy in predicting both the maximum and the total rate of erosion. ANFIS has recently been used in FDM to forecast different printing outputs based on printing parameters. FDM products' volumetric shrinkage was predicted by Dambatta et al. (Dambatta et al., 2019) using factors such as layer thickness, orientation angle, and structural geometry. Based on raster angle, layer height, and raster width, Rajpurohit and Dave (Rajpurohit & Dave, 2019) predicted the tensile strength of PLA printed parts using ANFIS. Yadav et al. (Yadav et al., 2020) investigated the impact of extrusion temperature, layer height, and material density on the tensile strength of PETG, ABS, and multi-material (60 percent ABS + 40 percent PETG) FDM products. They estimated the printed products' maximum tensile strength using ANFIS. According to our research, no significant studies have been done on how FDM-processed PLA parts react to slurry impacts. An ANFIS model is being developed to predict the slurry impact of FDM-processed PLA output as part of the current study to fill this void. Layer thickness, building orientation, and the slurry particle impact angle on the printed product's target surface are the printing parameters that are discussed.

## Evolutionary Algorithms (EA)

EAs are widely used in the AM domain to solve complex multiobjective design, process planning, and machine setup problems. Future applications are expected to benefit from increased computational power, allowing for more sophisticated and precise new algorithms. The increased exploration of EAs in the AM domain is sparked by the complexity created by combining additive and subtractive technologies. Finally, machine learning-based closed-loop process control and optimization could be critical in the drive to industrialize additive manufacturing. If additive manufacturing is going to be widely adopted in manufacturing, it must produce high-quality, repeatable parts. Engineering analysis (EA) tools are critical in the optimization processes required to achieve this goal, with collaboration between academics and industry serving as the final link in the chain.

## SUMMARY

Additive manufacturing stands to benefit greatly from artificial intelligence (AI) implementations. AI applications in streamlining additive manufacturing to be integrated into other manufacturing techniques or become a commodity for users are still far in the future, despite some progress having been made in this area. Design correlation, design improvement, defect reduction, and microstructural design are areas in which AI can help additive manufacturing. However, the main hurdle at the moment is the availability and reliability of data that are required for training. AM can significantly benefit from current approaches developed for other processes; the available experimental data has a wide range and is not always open to the public, thanks to the research community or AM manufacturers. In the development of AM-specific AI algorithms, ensuring the integrity of the data is critical. Experiments have been conducted or are currently being conducted with wide variations in observation and results. Creating a data storage facility is therefore critical for the manufacturing industry. The data generation condition must be disclosed along with the data to work correctly for the AI algorithms. Information such as process parameters, exact details of raw materials such as powder or feedstock, composition, raw material properties such as flowability of powder, particle size, etc., and the type of machine should be disclosed and shared to recreate label the data. In addition, the process's difficulties, such as high temperatures or high speeds, make it challenging to monitor and measure. The majority of currently available technologies rely on imaging the material's surface thermally or optically, and depth information is rarely readily available. Companies and academics must work together to create tools that accurately and quickly track a wide range of process conditions and parameters. For example, it's not yet possible to detect defects, create a 3D image of the build while it's being built, or keep track of the microstructure and grain orientations. These areas would undoubtedly present some difficulties that a willing and motivated community can overcome.

## REFERENCES

Anusuya, M. A., & Katti, S. K. (2010). Speech recognition by machine, a review. arXiv:1001.2267.

Band, S. S., Malvandi, A., Karimipour, A., Goodarzi, M., Afrand, M., Petkovi'c, D., Dahari, M., & Mahmoodian, N. (2015). Performance investigation of micro- and nano-sized particle erosion in a 90o elbow using an ANFIS model. *Powder Technology*, *284*, 336–343. doi:10.1016/j.powtec.2015.06.073

Bin Maidin, S., Campbell, I., & Pei, E. (2012). Development of a design feature database to support design for additive manufacturing. *Assembly Automation*, *32*(3), 235–244. doi:10.1108/01445151211244375

Dambatta, Y. S., Sarhan, A. A. D., Maher, I., & Hourmand, M. (2019). Volumetric shrinkage prediction in fused deposition modelling process–ANFIS modelling approach. *International Journal of Materials & Product Technology*, *59*(4), 347–365. doi:10.1504/IJMPT.2019.104568

Devlin, J., Chang, M., Lee, K., & Toutanova, K. (2018). *Bert: pre-training of deep bidirectional transformers for language understanding*. arXiv:1810.04805.

Everton, S. K., Hirsch, M., Stavroulakis, P. I., Leach, R. K., & Clare, A. T. (2016). Review of in-situ process monitoring and in-situ metrology for metal additive manufacturing. *Materials & Design*, *95*, 431–445. doi:10.1016/j.matdes.2016.01.099

Gu, G. X., Chen, C. T., & Buehler, M. J. (2018). De novo composite design based on machine learning algorithm. *Extreme Mechanics Letters*, *18*, 19–28. doi:10.1016/j.eml.2017.10.001

Hassan, M., El-Sharief, M., Aboul-Kasem, A., Singhbe, R., & Purbolaksono, J. (2012). A fuzzy model for evaluation and prediction of slurry erosion of 5127 steels. *Materials & Design*, *39*, 186–191. doi:10.1016/j.matdes.2012.02.012

Krizhevsky, A., Sutskever, I., & Hinton, G. E. (2012). ImageNet classification with deep convolutional neural networks. *Proceedings of Neural Information Processing Systems 2012,* 1097-1105.

Ladani. (2021). Article. *Journal of Physics: Materials*, *4*(4).

LeCun, Y., Bengio, Y., & Hinton, G. (2015). Deep learning. *Nature, 521*(7553), 436–444. doi:10.1038/nature14539 PMID:26017442

Li, S., Hassanin, H., Attallah, M. M., Adkins, N. J. E., & Essa, K. (2016). The development of TiNi-based negative Poisson's ratio structure using selective laser melting. *Acta Materialia, 105*, 75–83. doi:10.1016/j.actamat.2015.12.017

Rajpurohit, S. R., & Dave, H. K. (2019). *Prediction and Optimization of Tensile Strength in FDM Based 3D Printing Using ANFIS*. Springer.

Ramesh, C. S., Jain, V. K. S., Keshavamurthy, R., Khan, Z. A., & Hadfield, M. (2013). Prediction of slurry erosive wear behaviour of Al6061 alloy using a fuzzy logic approach. *WIT Trans. Eng. Sci., 1*, 109–119. doi:10.2495/SECM130091

Savkovic, B., Kovac, P., Dudi'c, B., Rodi'c, D., Taric, M., & Gregus, M. (2019). Application of an adaptive "neuro-fuzzy" inference system in modeling cutting temperature during hard turning. *Applied Sciences, 9*(18), 3739. doi:10.3390/app9183739

Shevchik, S. A., Kenel, C., Leinenbach, C., & Wasmer, K. (2018). Acoustic emission for in situ quality monitoring in additive manufacturing using spectral convolutional neural networks. *Additive Manufacturing, 21*, 598–604. doi:10.1016/j.addma.2017.11.012

Shi, Y., Zhang, Y., Baek, S., De Backer, W., & Harik, R. (2018). Manufacturability analysis for additive manufacturing using a novel feature recognition technique. *Computer-Aided Design and Applications, 15*(6), 941–952. doi:10.1080/16864360.2018.1462574

Wasmer, K., Le-Quang, T., Meylan, B., & Shevchik, S. A. (2019). In Situ Quality Monitoring in AM Using Acoustic Emission: A Reinforcement Learning Approach. *Journal of Materials Engineering and Performance, 28*(2), 666–672. doi:10.100711665-018-3690-2

Yadav, D., Chhabra, D., Gupta, R. K., Phogat, A., & Ahlawat, A. (2020). Modeling and analysis of significant process parameters of FDM 3D printer using ANFIS. *Materials Today: Proceedings, 21*, 1592–1604. doi:10.1016/j.matpr.2019.11.227

Yao, X., Moon, S. K., & Bi, G. (2017). A hybrid machine learning approach for additive manufacturing design feature recommendation. *Rapid Prototyping Journal, 23*(6), 983–997. doi:10.1108/RPJ-03-2016-0041

Zhang, X., Le, X., Panotopoulou, A., Whiting, E., & Wang, C. C. L. (2015). Perceptual models of preference in 3D printing direction. *ACM Transactions on Graphics, 34*, 1–12. doi:10.1145/2768821

# Chapter 1
# Application of Machine Learning Techniques in Additive Manufacturing: A Review

**Amithkumar Gajakosh**
*BTL Institute of Technology, India*

**R. Suresh Kumar**
*B.M.S. College of Engineering, India*

**V. Mohanavel**
*Bharath Institute of Higher Education and Research, India*

**Ragavanantham Shanmugam**
*Navajo Technical University, USA*

**Monsuru Ramoni**
*Industrial Engineering, School of Engineering, Math, and Technology, USA*

## ABSTRACT

*This chapter provides an analysis of the state-of the-art in ML applications for optimizing the additive manufacturing process. This chapter primarily presents a review of the literature on the use of machine learning (ML) in optimizing the additive manufacturing process at various stages. The chapter identifies ML-researched areas in which ML can be used to optimize processes such as process design, process plan and control, process monitoring, quality enhancement of additively manufactured products, and so on. In addition, general literature on the intersection of additive manufacturing and machine learning will be presented. The benefits and drawbacks of ML for additive manufacturing will be discussed, as well as existing obstacles that are currently limiting applications.*

DOI: 10.4018/978-1-7998-8516-0.ch001

## INTRODUCTION

We've been curious about programming computers to learn since their inception. Having a firm grasp on how to programme and teach machines to learn based on historical data and current experience is critical when making decisions. When it comes to Additive Manufacturing, where quality is constantly changing, consider using machine learning techniques. Consider machine learning as a new way of seeing additive manufacturing. Additive Manufacturing (AM) process environment obstacles, such as monitoring AM materials and parts, are a common theme in modern AM data. Machine Learning for Additive Manufacturing algorithms can help transform and power output for better quality, lower costs, and process optimization.

The quality of printed products is dependent on a variety of process parameters, which is one of the major challenges of AM (also referred to as 3D printing) (speed and thickness). All of these links are illogical. A powerful and efficient data mining and data analysis tool is required to meet these challenges. As a result, we use a branch of artificial intelligence called machine learning to deal with these problems. With machine learning models, optimal process parameters can be predicted and determined based on the training data set and the lessons learned. It finds connections and connections between things. Design professionals use it the most, as well as those involved in AM production and the AM process itself. By combining different materials, additive manufacturing can create a wide range of different parts. Additive manufacturing can create even complex geometric structures in a single step, whereas traditional methods necessitate more steps (Razvi et al., 2019; Wang et al., 2020; Witherell, 2018). In the figure 1, can see the AM lifecycle from Design to production process.

## FUSED DEPOSITION MODELLING (FDM)

The additive manufacturing process involves creating a CAD model, converting it to the required format, and feeding the converted file format into a 3D printer. A layer-by-by-layer printing procedure is used in additive manufacturing (AM). The AM process can be broken down into seven different types according on the ISO/ASTM52900-15 standard. FDM is classified as a Material Extrusion process since it uses polymers as a feedstock to produce finished items (Keshavamurthy, Vijay, & Saravanabavan, n.d.; Sood et al., 2013; Wang et al., 2017a).

Polylactic acid, acrylonitrile butadiene styrene, and polycarbonate are some of the materials often used in FDM applications. These polymers are transformed into 1.75mm or 2.85mm filaments before being fed into FDM machines, depending on the requirements. Using thc appropriate temperature, the filament is fed through

*Figure 1. AM lifecycle, examples of associated data, and decision making applications*

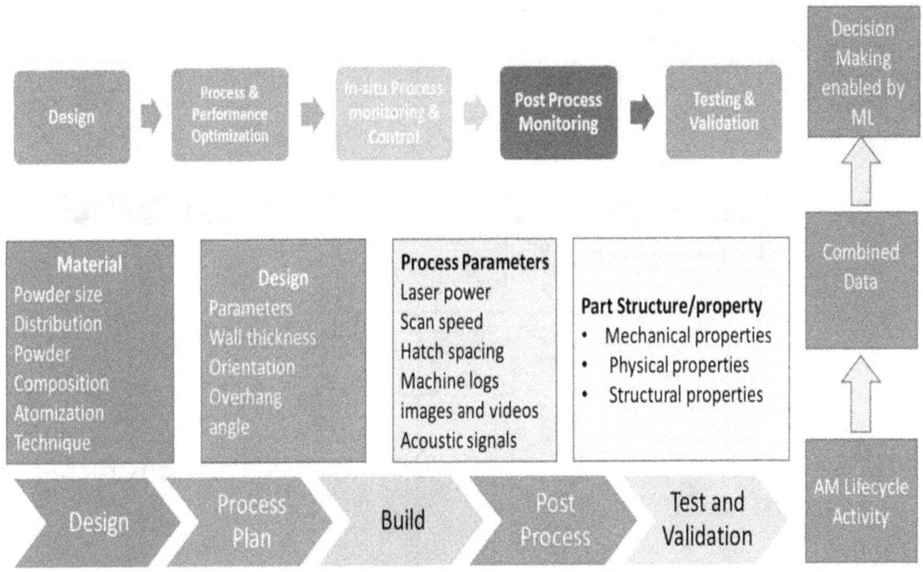


the nozzle and deposits a layer on the print bed after changing phases. A solid part is created by stacking the remaining layers, bonding them, and then cooling them. Figure 2 (Tambrallimath et al., 2019) shows the FDM schematic diagram. The typical qualities of a material are determined by a number of process parameters. By adjusting these settings, the material's desirable qualities can be improved. It's possible to alter settings like raster angle, orientation, bed temperature; nozzle temperature, raster width, and infill to get the results you want (Tambrallimath et al., n.d.). Diverse materials are being investigated for use in FDM by a number of scientists. Vijay et al. produced and developed a PC-ABS+Graphene polymer nanocomposite by altering the graphene content. In FDM, the combination of these materials has led to effective part development and to their utilization. With no evident faults, the created filament showed how well it could be used in FDM (Keshavamurthy, Tambrallimath, Ugrasen et al, n.d.). A novel composite was created by Fuda Ning et al. (Ning et al., 2015), who used FDM to test the tensile and flexural properties of ABS thermoplastic reinforced with carbon fibres. At a rate of 2m/min and a diameter of 2.85mm, the produced composite was extruded as a filament at 220þC from a through-nozzle. ASTM standards were followed when conducting the tests. Both the tensile and flexural characteristics of the material improved simultaneously. The tensile and flexural characteristics of FDM-developed polymer composites were studied by Jaya et al. (Jaya Christiyan et al., 2016). Hydrous magnesium silicate was used to strengthen ABS instead of just using plain old plastic. The qualities were examined

*Figure 2. Schematic of FDM (Tambrallimath et al., 2019)*

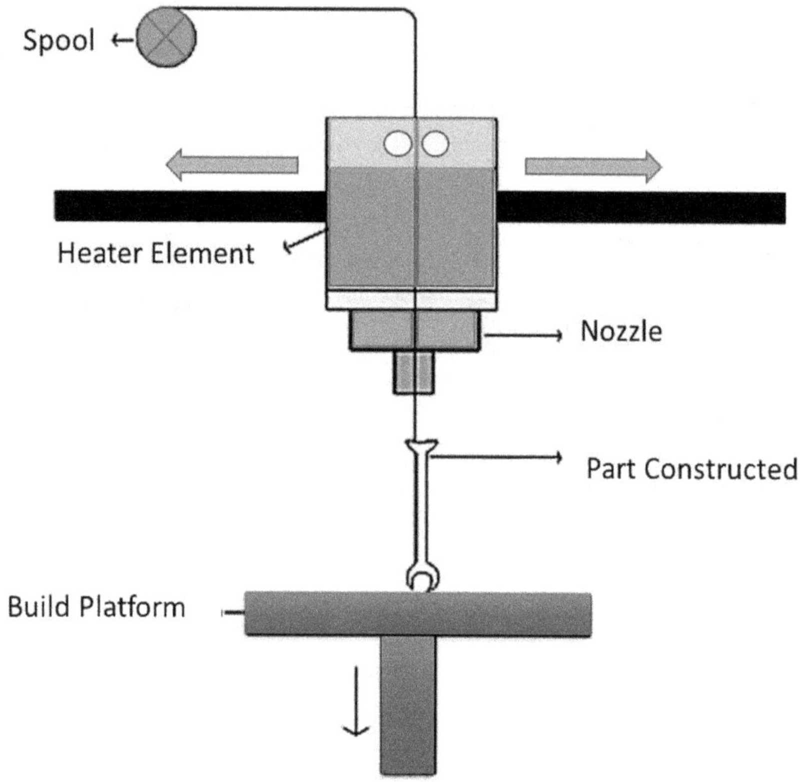


in accordance with ASTM guidelines. Printing speed and layer thickness were both investigated as possible causes of property variation. The best tensile and flexural characteristics were found when printing speed and layer thickness were reduced.

Additive Manufacturing is embracing machine learning, a relatively new area of study. Data analysis or prior findings are used by machine learning to help optimize desired outcomes. Prediction, optimization, flaw detection, regression, and data classification are all possible uses for machine learning. The data is used to train the model. The use of machine learning in additive manufacturing has lately grown in popularity, enabling numerous researchers to improve product quality, optimize process parameters, and reduce costs. AM-based product development comes with its own set of limitations, such as inconsistent fabrication parts that are highly dependent on a variety of process parameters. ML principles introduced to the FDM process are covered in the following sections of this chapter.

## MACHINE LEARNING ISSUES

The amount of training cases, issue representation or feature extraction, and self-learning affect accuracy in learning systems, as does when and how algorithms should be utilized for approximation. It is curious about the theoretical limits of comprehension. It's not uncommon for machine learning to run into scaling challenges due to factors such as the amount of data input and output, as well as training vs. testing. In some cases, machine learning (ML) performance seems challenging for human professionals. Human learning ability may be superior, but it cannot be equaled by artificial intelligence (AI). When it comes to the reasons why machine learning succeeds or fails, we know very little. Changes in symbols rather than numerical information are being credited with the achievement's success. ML is already in use in a wide range of fields, including finance, healthcare, and retail. ML has been compared to a neural network's brain, yet it is fully guided by humans. Figure 3 shows the ML and algorithm structure.

## MACHINE LEARNING (ML) FOR FDM-AN OVERVIEW

When it comes to AM methods, a useful tool for quick prototyping is fused deposition modeling. A nozzle compacts filaments in a platform to build layers on a thermoplastic substance. Energy assessment efficiency is one of the essential study issues. Simulated FDM isn't a possibility at the moment; hence the focus of this chapter is on FDM machine learning.

According to the literature review, ML was used to characterize melting mechanisms, identify flaws, and analyze product shape when applied to the AM factor.

- For AM techniques alike, the prediction strategy can reduce energy usage by as much as 50 percent.
- It was decided to employ the random forest technique to emulate FDM capacity. When structured data is provided, it is possible to utilize a random forest technique that is faster than the alternatives. A random forest can be created from many decision trees, boosting precision while reducing over fitting (In ML over fitting means model matching the exact data). Repeated layers of material had to be overcome during the FDM step.
- In a true coded genetic algorithm, all decision variables are coded as rational data except for the final stages, which resemble binary-coded genetic algorithms. Generating real-life decision factors completely solves the problem, as opposed to binary genetic coding methods, which present numerous challenges (e.g. constant decreasing of search areas, unable to

*Figure 3. Structure of ML techniques and algorithms*

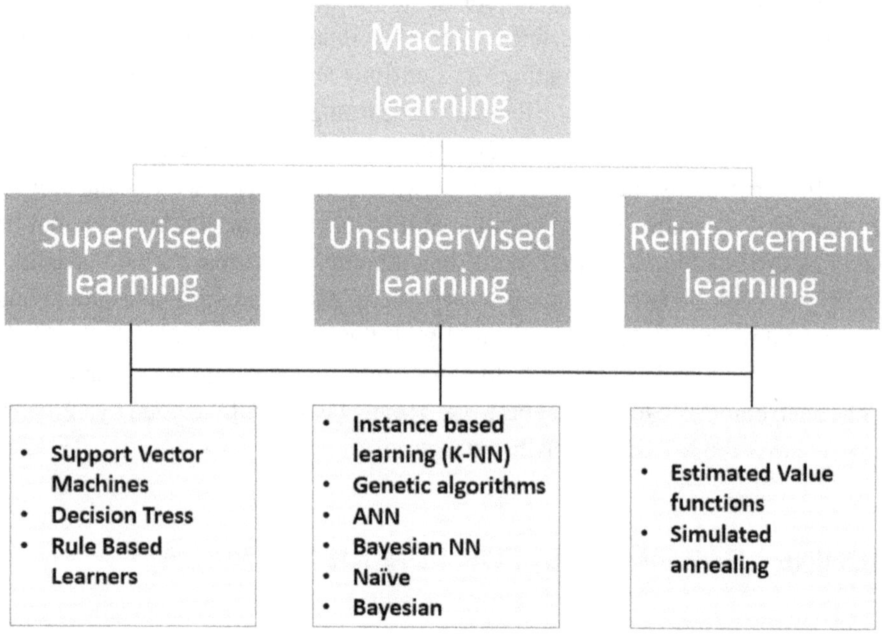


obtain arbitrary response accuracy, fixed mapping of problem variables, etc.) It's suitable to use coded GA for this journey because there are infinitely many possible outcomes. The problem must be complicated as well due to the part's textured surface. Mathematical functions used to define a problem's aim may evolve in tandem with geometry since geometry is constantly evolving. GA approaches are the ideal approach to solving this problem due to the fact that it is multi-modal. Using the created vectors and rotation angles in the population as a starting point, the revised population is then used to generate new vectors and rotation angles. The coordinates are 0þ North, 1þ South, 1þ East, and 0/360þ. To align the nx, ny, and nz vector axes, the amplitude of the unit vector is employed. First, we need to obtain the weight component formulas W1 and W2, and then utilize that equation to solve for O to find the objective value. It used to be that the value of the two objective functions was expressed using both and. This problem uses the selected pressure of two, with the selected pressure being equal to the total number of solutions in the event of selecting a tournament. About 75% of the time, the two-step crossover and transmission operators will have a probability of 25% or below. Two-step crossover and transmission operators have a 75% or higher likelihood in most circumstances. The software is used to find the solution for the user number. Parts that are fully free to fit can benefit from this strategy, regardless of how simple or complex they are. This strategy, on the other hand, may be inefficient in terms of calculation for simple bits. Several GA parameters are fed into the device, as well as the STL part register and the minimum and maximum cut thicknesses that must be maintained at all times on Ra. The strategy described above is verified by comparing its predictions with those found in the literature (Phatak & Pande, 2012). There are working examples of the framework available. The creation and implementation of a system one step at a time is taken to address the many aspects of framework architecture, development, and testing.

## Modular Design

Users provide STL CAD component representations as part of their design processes. An OOP-based hierarchical data structure like the aerial edge data structure is created by reading and processing STL data. The genetic algorithm-based method is used to find the best computational path. An external outline was built to construct an offset 2D slice plane internal contour with the correct breadth using geometric progressions put down on paper. The amount of material used in the guided hollow CAD model is counted. As well as time and expenses, it considers the position of each component, the quality of construction, and any step errors, as well as the volume and area of support structures that may be necessary.

Listed below are numerous system components that will be discussed in great detail in the sections that follow.

## Preprocessing Module

The device is fed by the STL CAD model's portion. Aerial Edge, an OOP data system, was created and built to update STL data. Nodes (vertices), boundaries, faces, and connection are all shown in Figure 3. Because to the improved pointer arrangement, the search difficulties have been reduced. There are no double entries in the data structure to improve speed and data integrity. Dynamic data formats like the vector list are employed for efficient memory storage. Both succeeding modules are connected to the OOP structure that shares data.

## Part Orientation by Means of Genetic Algorithm

It governs the production capacity and output for each component in the RP phase as well as the time and output specifications for each step. Binary coded GA rather than typical gradient-based optimization approaches were used in the current study to achieve optimal component orientation. If you're looking for a way to discretize search space using binary coding, then binary coded GAs are best bets. Finite-length strings for real variables pose several obstacles, including as the inability to achieve solutions with arbitrary accuracy, the mapping of fixed problems, and a difficulty with Hamming cliff inherently. According to this argument, real-coded GA crossover operators cannot do constant search space searches (such as in constant-space search). In this case, binary-coded Gaussian-A may be used; however this problem now occurs since the search zone can support an infinite number of orientations. The solution used to calculate the part's parameters is heavily influenced by the surface complexity. Quantifying the volume of structures that come into touch with an object is a frequent method employed in these projects. The weighted mean of these five is calculated using functional data, and GA is applied to reduce the weighted mean.

## Chromosome Structure

The Y axis is used to direct movement. The Z axis points in the direction of construction on a CAD model, whereas the X and Y axes rotate the model. For best orientation, consider the relative location of x and y as a variable because they rotate at an angle to one another. The difficulty has been noticed in writing, and one suggested answer is the chromosome. X and Y axes make form a binary digit string; each chromosome represents an allowed orientation for these various components. x and y have bit string lengths of 16 bits each, corresponding to a 0 to 180-piece range. Thus, a chromosome contains 32 bits of information. Table 1 shows a typical x-bit DNA string, which is a partial chromosomal structure. A huge number of chromosomes make form a species. When a population grows, scientists try to see how big they can make it. These random values were created randomly within the specified range in order to create fault and calculate digital consumer factor for x, y and build-height error using RP method and then calculate digital consumer factor for these values.

## Crossover and Mutation

It is through proportional selection that chromosomes are paired to generate whole genomes. Chromosome pairs include duplicate chromosomes, each of which has one of its kinds in the appropriate proportion to physical fitness. The sorting approach is used to merge chromosomes in the same pool when chromosomal fitness is evaluated descendingly. In general aviation literature, there are a plethora of crossover strategies documented. Because of this, all previous configurations have been put to the test (such as a one- or two-point crossover, or a uniform crossover). One- and two-point crossovers were proposed for a larger crossover with a very wide search space, and that was the typical crossover. To further analyze the search region, researchers created a single-point crossover method, which was applied in this study. Table 1 shows how the chromosomes of the parents (P1, P2) are separated into two pieces (x chromosomes) and passed on to the children (C1, C2) for crossover. As a result of the single-point Crossover operation with x = 0, the following children were generated: Sections of parent chromosomes light up, and parts of child chromosomes are exchanged, as seen in Table 1. More than one test to locate the most successful method of searching has showed that the determining level is higher and the mutation probability is below the threshold, respectively. This did not mean that there were more advantages because mutation rates were higher.

## The Fitness Function

The evolutionary algorithm requires a single value in the goal function when attempting to find a feasible solution (orientation of the part). Scientists have applied efficiency tactics to boost part orientation in fitness, such as expanding size, maintaining consistent, establishing stability, and lowering package volume or area, according to articulated literature. Even while no attempt has been made to further integrate the material use into the fitness role, in particular, to a hollowing of the parts, there have been few attempts to change the fitness operations to better take use of this usage. Ultimately, the goal of the research is to create a new hybrid fitness factor by making the greatest possible use of all of the different components. Our methodology is based on a weighted average, which takes into account all five of the criteria. Using a system that integrates w1 for building height, w2 for the step error factor, w3 for material usage, w4 for the element surface in touch with support structures, and the capacity of the supporting structures, the user establishes weights for the five factors, one of which is step error. The importance of a user specification to the RP process and other user requirements varies, of course. Stereo lithography (SLA), FDM, and selective laser sintering can all be used with the approach given here (SLS). Weighting factors can be used to define the relative importance and weight of certain criteria. This is critical because it makes it easier to make changes and adjustments to the overall production environment while the project is still in progress. A population's optimal (x and y) orientation is determined by an iterative optimization process in the GA module. Finding the optimum solution involves finding the one with the highest fitness function value and recording the right orientation angles (x, y). Eq.1 recommends a health function (F) that is on the low end of the scale.

*Table 1. a) Partial chromosome (θx). b) Crossover operation.*

| 1 | 0 | 1 | 1 | 1 | 0 | 1 | 0 | 1 | 1 | 1 | 0 | 0 | 1 | 1 | 0 | |
|---|---|---|---|---|---|---|---|---|---|---|---|---|---|---|---|---|
| (a) | | | | | | | | | | | | | | | | |
| P1 | 0 | 1 | 0 | 1 | 1 | 0 | 1 | 0 | 1 | 1 | 1 | 0 | 0 | 1 | 1 | 0 |
| P2 | 0 | 0 | 1 | 1 | 0 | 1 | 0 | 0 | 1 | 0 | 1 | 1 | 1 | 0 | 1 | 0 |
| C1 | 0 | 1 | 0 | 1 | 1 | 0 | 1 | 0 | 1 | 1 | 1 | 1 | 1 | 0 | 1 | 0 |
| C2 | 0 | 0 | 1 | 1 | 0 | 1 | 0 | 0 | 1 | 0 | 1 | 0 | 0 | 1 | 1 | 0 |
| (b) | | | | | | | | | | | | | | | | |

$$\text{Maximize } F = \frac{1}{1+(\text{w1 x H})+(\text{w2 x Ra.avg})+(\text{w3 x PM})+(\text{w4 x A})+(w5\,x\,V)} ---(1).$$

If standard performance criteria apply, the true construction height, H, will be higher. RP hollowed section material use factor is PM (building staircase error), wherein A (supporting surface contact area) appears to be the enabling surface contact area and V (support volume) is the support volume. Ra is the staircase error factor.

$$\sum_{i=1}^{5} Wi = 1.0$$

## Build Height

Component model orientation for which bounding box sizes (x-size, y-size, and z-size) are required should be selected before computing the bounding box size. The bounding box's size can be used to calculate the diagonal. A scaling factor of 1.2 was used to maintain consistency in the bounding box size. Using Eq.2, we can figure out how high the building must be to meet code.

$$\text{H} = \frac{\text{Z}\max - \text{Z}\min}{1.2\text{ x diagonal of bounding box}} ---(2).$$

Where $\text{Z}\max - \text{Z}\min$ represents the actual oriented part build height along Z axis. Staircase error factor (Ra.avg) is calculated as follows:

$$\text{Ra.avg} = \frac{\sum Ra_i A_i}{\sum A_i} --- (3)$$

$$\text{Ra}_i(\mu\text{m}) = \frac{\text{slice thickness}(\text{mm})\,\text{x}\,70.82}{\cos,} --- (4)$$

These values are explained by the relationships shown below. Ra avg is the average staircase factor for the part Eq.3.Roughness Eq. 4 Rai, as well as the region of its triangle STL file $i^{th}$ and the angle between the regular face's regular face and the Z-axis, are all important variables. Part manufacture does not require any support structures when using the SLS method. It's not significant in this work and thus isn't taken into account for the current issue because it's concerned with support (A) as well as the number of support systems (V). Even though processes like SLA or FDM necessitate a computation, the approach should nonetheless offer it.

The surface roughness in FDM was predicted using feed forward neural networks, and the evaluation function devised to select the optimal solution was employed (Boschetto et al., 2013).

For the purpose of determining the resilience of the FDM process across the deposition angle, a neural network will be constructed. This neural network will be used in the experiment. If you're familiar with the term "neuron in the brain," you know that an ANN is a network of individual nodes connected by a variety of linkages, also known as linear and nonlinear scalar transforms. Using empirical data as input, neural networks are capable of performing a wide range of functions such as categorizing, estimating, modeling, and forecasting A low-order empirical polynomial method may prove inadequate if there is no acceptable analytical model. In this case, neural networks can be used.

The ability to tolerate noise or partial data tolerance, several variables accommodation, and unexplained interactions are all desired features in addition to those already mentioned. Several studies have been conducted on the prediction of surface roughness during machining processes and published in peer-reviewed journals. It's possible to classify them as either conventional or non-conventional in nature. First and first, a great deal of work is put into performing changes like washing, frying, and boiling. With the help of algorithms for calculating surface roughness, the ANN established a full link between cutting parameters and roughness. Several pieces have been created using an on-the-fly approach to monitor and correct for machine uniformity. Instead of assessing surface roughness, the form of the performance variable should be in the shape of a flowing figure that shows a variety of outcomes. Off-line replication of network models was a significant objective of the research. A precise description of surface roughness is necessary for the strategy to work, according to previous research. Any non-contact or non-invasive action is also rejected. A component's and a sheet's thickness can both be changed in FDM, however in order for the component and sheet thickness to remain predefined during the machining process, FDM requires both their axes to be established ahead of time. For example, the electrical discharge machine and rapid prototyping (RP) both used ANN technology. The author created a neural network technique that simulates surface roughness using an artificial neural network to imitate the information flow. Using a programme named Raster2Polygon, it analyses five different input factors, such as the thickness of the sheet and the direction of the raster. Consequently, the random sequence generator converged faster than the normal random generator, showing that the chaotic random numbers generated were better than the random numbers in the order in which they were generated. However, only vertically and horizontally-oriented surfaces are used to repeat the roughness measurements that are restricted to the region of Ra. As previously indicated, only two factors in the depositing angle range of 45–135 are significant during the FDM phase. Furthermore, the change in performance for the remaining variables is negligible. Stratification direction (and hence which factors affect it) and the amount of each product required are all that operators in the FDM production area need to know ahead of time, according to the authors. Model filling and help creation are examples of additional factors that can be altered after model construction and prediction. This model was chosen because of its ability to accurately predict surface roughness, which is heavily influenced by the deposition angle and sheet thickness. The author believes that the first parameter should be carefully addressed because many events appear to occur and interact with one another at exact inclinations, resulting in a large number of cases being presented to the network. This particular challenge was carried out by a traditional artificial neural network. An ANN's basic building blocks are neurons, weight, and training rules. Essentially, every neuron is a simple processor that receives input signals and processes them to produce an output that is then supplied as an input to another neuron for processing. The input layer, the hidden layer, and the output layer are all parts of a neural network. The network inputs are everything that is in the input layer. At least two secret levels follow, with many parallel neurons along the walls of each. Neuron function is the result of performing a weighed summation of the inputs in each neuron. After that, the neuron function's output is sent into an output layer that uses a nonlinear activation function to create the desired effect. With the training algorithm, the weights and biases of the network are dynamically adjusted so that the intended error between current and target outputs is reduced.

Neural networks perform, efficiently, and effectively depend on network characteristics such as the number of training data, learning rate, number of hidden layers, and processing functions used. However, there are no hard and fast rules for determining the number of neurons in a hidden layer or how many hidden layers there should be in a pyramid. The training package will take longer to complete as the number of layers grows. This usually starts with an OHL layer, which can be buried or not. For example, sophisticated learning models and reinforcement learning use highly specialized modeling that is considerably more restrictive, whereas common models are more generic. Multi-layer perceptron networks (also known as neural feed-forward networks) appear to have a non-linear functional activation structure. Although neurons within the same layer were not connected to incoming connections, they are highly intertwined. Input and output signals are routed through the hidden layers such that they only travel in one direction. A FF network with x1,...,xn hidden layers and two visible layers available. This is how the algorithm works: As can be seen from Eq.5, the method is dependent on a hidden neuron (Wu et al., 2015; Wu et al., 2017).

$$\sigma\left(\sum_{j=1}^{n} w_j x_j + b_j\right) - - - (5)$$

As you can see in equation, where wj, bj is represented by the weights with the neuron arrows, the connections are as shown. Eq.6 dictates the calculation of the network output, which is equal to the weighted sum of an output units in the hidden layer.

$$y = \left(f \sum_{i=1}^{n} w_i x_i\right) - - - (6)$$

According to this theory, to produce precise output patterns in response to specified input patterns, weight connections need to be changed appropriately. To put it another way, we can train a neural network to perform a specific function by varying the strength of the neural connections. A network's performance is evaluated when any new additions are made to check if they satisfy the aim. The difference between the source and the network output is used to calculate the error. The network's goal is to have a low average number of these errors. Some prototype models have been constructed to collect data for use in training and validating ANNs as part of an experimental (pilot project). Creating a neural network is just the beginning; the real work begins after that. Rather of treating each move as a single process, the focus was on ensuring repeatability and reliability. It was decided to use cylinder-shaped samples with coordinates perpendicular to the development axis and generation perpendicular to deposition axis, in order to keep the angle fluctuating throughout time. The diameter was increased to 100 mm since more research is needed in specific areas of the study. CAD/CAM modeling and translation have a problem because they are not robust enough to guarantee an error-free product. CAD/CAM modeling is not used in the Wolfram Mathematical platform's virtual model, which is created by hand using CAD. In this case, the 3D modeler-created geometry was employed in the Mathematica environment, where unwanted mesh anisotropy was obvious. There are no issues with this particular STL. The sample was designed on a Stratasys BST 768 Dimension, which has absolute dimensions guaranteed at 0.127 mm. The whole room height ranges from 203 millimeters to 305 millimetres. For the models and supports, it is said that they are manufactured of ABS, with tensile qualities of 22 MPa, stress modules of 1.6 GPS and glass transition temperature of 104 degrees Celsius (F). The chambers and nozzle are set to a temperature of 75 degrees Celsius and 270 degrees Celsius, respectively. Model filling was scant due to the usage of two different segments of the same layer of thread (0.254 & 0.331 mm). There was also a revelation in regards to the carrier's technique: multiple parameters, some of which are critical to surface uniformity, have an impact on the sensitive areas (Liu et al., 2017). An external validation of specimens was performed to test the outcomes of several procedural parameters. It was determined how much a filling method (whether dense or sparse) and the supporting strategy differed, as well as the thickness of these layers (surround and break). It's possible that the addition of a second verification will help determine how long the net model will last, given the technology and equipment used. For this application, we looked into ULTEM 9085 along with ABS plus and Polycarbonate as three other options for plastics. Two Stratasys materials were used to 3D print this part; a layer with a thickness of 0.254 millimetres was created on each one. Stratasys applied a 0.254 mm layer of thickness to the Fortus 360's prior layer before it was finished off by the user.

## HYBRID MODELS

Incorporated with the low-square matrix genetic algorithm. Fused Molding Component Wear Strength is Predicted Using a Model Integrating Wear Testing and Design (FDM). The whole structure, as well as the resources required to build it, had already been meticulously planned before building began. Layer width, orientation, raster angle, raster width, and air gap were all part of the technique. Researchers found that utilizing a combination of ML algorithms was more effective than using a single algorithm based on the values of these parameters, which were supported by GP and NN input.

Simulating FDM wear strength has been done using GP and ANN ensembles. Genetic algorithms (GAs) use Darwin's theory of evolution to describe how evolution works in nature. This is the foundation of GP. GP applications may turn up new material property models. One technique to solve a problem is to use GPA, but in GGA, the solution is expressed as text or as a binary value instead of a tree structure. In the GP algorithm, it all starts with creating the first human beings. The population's size is determined by the number of people born. In other words, everyone is a mix of functional and terminal components. The terminal components are used to create constants and troublesome input process parameters. This research looked at five different stages of the input process. Each individual solution is assessed based on fitness functions such as the root mean square error once a set of starting answers has been established (RMSE). Until the termination conditions are met, genetic modification techniques such as selection, cross-section, and mutation will continue to evolve and be used. The criterion for terminating the simulation might be the number of generations specified by the user, or it could be the model's error threshold. Using a 200-parameter set and a 300-generation-number generation, this study explores the effects of crossover, transformation, and selection using parameters such as 0.85, 10 and 0.05 that were determined through a process of trial and error. ANNs are currently considered one of the most promising ways for building non-linear models purely based on data Input, hidden layer, and output layer make up the three levels of ANN structure. There are neurons in every layer. During system training, weights are selected to connect input data with the hidden layer. The input layers are given weights based on the amount of data they contain to produce weighted sums. For each hidden and output layer function, a corresponding transfer weight is applied; this yields a final result that can be determined. The output is constructed by mixing the weighted average of the hidden units with the activation function results using the transfer function (usually the sigmoid). Nonlinear relationships are often modeled using the sigmoid function, which has a range of -1 to +1. Another feature of ANN networks is that they constantly update their weights, so that there is less of a discrepancy between their real-world output and their network-generated one. Recurrent propagation is a common training method for ANNs. This method redistributes the model's output and then tweaks the model's weights to reflect the new distribution. Numerous iterations are needed to get to the point where the quantity of error is as low as possible. It uses a second derivatives-based approach to maximize the weights and hence achieves a faster convergence timeframe. In order to determine the right rpm setting, we first use training data to estimate the best hidden neurons in hidden layers. The RMSE for neural network three is again confirmed to be good based on previous tests. As a result, in our research, we've opted for a neural net with three hidden neurons, most of which are located on discrete units. The GP and ANN approaches are used to represent an FDM process in the current work. A low-squared ensemble model is constructed by combining the GP and ANN forecasts using the linear low-squared technique (Garg & Tai, 2014; Zhu et al., 2018).

## ARTIFICIAL NEURAL NETWORK (ANN) FOR AM PRODUCTION PROBLEMS

In this method, the first step is to choose a training experience, then the goal function, its representation, the function approximation procedure (estimate training values, change weights), and lastly the final design. Data is used to power neural networks. The amount of data that can be accessible has a direct impact on performance. In certain areas, they've created their own datasets, but in AM, no huge datasets are accessible because the cost of training data is so high. This means that we'll have to start from scratch and create our own open-source data set from scratch. Small datasets are therefore crucial. We can use advanced methods like generative models to generate augmented input for our datasets.

Surveillance learning is commonly used in neural networks since it requires the output of a target. We need to understand how to link the numerous 3D printing parameters' attributes together. Only by working together closely with professionals in computer science and material science have these results been made attainable.

Genetic programming and neural networks were merged with the regular least square method in a second way to reach this result. With this model, the wear resistance of FDM aerospace components may be calculated. There is a model for every single variable (FDM). Each component had a predetermined design and construction, as well as the raw materials employed. Thickness, orientation, raster angle, raster width, and air gap were all taken into account during the drafting process. This data was fed into the GP and the NN models. One machine learning method was shown to be statistically less accurate by the researchers (Yia et al., 2019) than a combined strategy. Machine learning has numerous uses in additive manufacturing.

## 1 AM LIFECYCLE

In order to classify a functional vector into a known class, the nearest neighbor technique is used. With this method, you can figure out how far away an unknown vector is from its neighbors. This classifier utilizes the Bayesian approach to decision-making, which is the foundation of the Bayes Classifier. Pattern recognition in a wide range of industries, such as picture and text recognition, uses this statistical methodology extensively. Neuroscientists frequently employ neural networks in their work. There are three levels to it, each with a different number of neurons. Neuronal networks can be trained using a wide variety of strategies.

In current machine learning, SVM is a statistical learning theorem-based methodology. In the realm of condition monitoring, it's likewise a well-liked technique.

## MACHINE LEARNING ALGORITHM ADVANTAGES

In order to address the same problem, both support vector machines (SVM) and decision trees employ different methodologies. There are certain advantages to using decision trees over support vector machines. As someone with a background in machine learning, Christopher analyses how machine learning may be used in manufacturing. Increase the length of the sentence: Furthermore, given current industrial concerns, machine learning applications may be preferable. A number of NP-complete problems that occur often in the manufacturing process have been successfully addressed by the machine learning sector. Numerous variables, including increased availability of large volumes of complicated data with limited transparency and rising application and strength of existing ML devices, have contributed to the increased adoption of ML approaches in recent decades.

In spite of this, the fundamental ML principle of allowing participants to solve issues without having to write code remains current and useful. Data mining, machine learning and artificial intelligence are all words that are regularly used in connection with ML. A wide range of industrial sectors, including optimization, administration, and troubleshooting have adopted machine learning.

There are many ML algorithms (such as a support vector machine) that are built to handle enormous datasets and manage high dimensionality. In addition, these authors write in another piece that concurrent difficulties such possible over fitting should be expected at all times during the application process. There are techniques to lower the dimensions so it is achievable even if the algorithms cannot handle it. Researchers assert that dimension reduction's influence on projected results has waned.

Because dimensions cannot exceed each other while using SVM, ML is essential. Furthermore, while diminishing dimensionality isn't necessary, the pressure to do so is relieved. For example, permitting the use of trivial production data to estimate business growth may become more permissive, which could be important in some circumstances. Once it closes the knowledge gap, this will be hugely advantageous.

By gathering, processing, and/or analyzing existing data sets, machine learning may be able to help extrapolate future trends and make predictions about the program's future behaviors. Process owners can utilize this additional information to make better decisions, or system components can use it directly to improve the system. Several machine learning methods are primarily concerned with evaluating patterns or regularities that characterize interactions, as previously indicated. To us, artificial intelligence and machine learning (as well as some NLP tools) are all parts of machine learning (ML), and that they can learn to "keep up with and include" changes in the dynamic production environment, regardless of what the system designer needs to plan for and offer solutions to. Since many first-party models struggle with adaptability, ML gives an excellent rationale for why ML industrial applications may be advantageous. The advantage of machine learning is that it automatically adjusts to changing situations and learns from them. Pre-existing data is mined using machine learning methods (ML). That recorded data is only useful if processed and turned into information that may be utilized to generate predictions, for example. In a live production programme, real-time data cannot be collected, and technological, budgetary, and knowledge limits further complicate matters. Manufacturing is particularly impeded by all of these reasons. These design factors may have an impact on the location of process control points. The drawbacks of picking checkpoints based on their importance can be overcome by using machine learning approaches that have analytical power and capability. Perhaps it no longer matters as much to collect information by repurposing old, irrelevant data. This could result in the discovery of new states and the collection of all production-related data. It's still up for grabs, so whether or not this is a good thing is anyone's guess. There are no technological obstacles in processing the new data because it is well-suited to ML's ability to handle high-dimensional information. The data may be difficult to obtain, especially if data capturing capacity is an issue. When data is made available, or if the requirement arises in some other way, the overarching problem of extracting state drivers from complicated contexts generally isn't seen as a challenge. With regard to various industrial applications' problems, the following picture shows the theoretical production capacity of machine learning (ML). This suggests that the abilities associated with intelligent creation are part of what we consider to be an individual's inherent intelligence. It appears that ML relies on solid arguments to disprove the limitations and difficulties associated with the conceptual product state paradigm. The above-mentioned analysis shows that ML approaches may be a good fit based on the needs. Due to the large majority of requirements being met, ML is quite beneficial.

ML approaches' strengths and limitations must be evaluated in greater depth in order to achieve the above-mentioned standards. In other words, the product state can only be enhanced by considering whether or not the theoretical notion of the product condition and the programme concept are met before a final decision is taken. Additionally, there are other questions about how ML algorithms respond to qualitative input. The next section analyses the advantages and disadvantages of machine learning in manufacturing, based on the criteria already established.

## Advantages and Problems of the Implementation of Machine Learning in Production

Various manufacturing process optimization, monitoring, and control applications, as well as several fields of predictive maintenance have demonstrated that machine learning is a significant tool. Although ML approaches have been proved to improve quality control optimization in production systems, they have not yet been shown to have a major impact on reducing issues in a wide range of contexts where it is difficult to identify the root causes of errors. But the vast majority of machine learning (ML) systems are employed for discrete tasks, not for the entire production or industrial system. There are a wide range of ML methodologies, tools, and methods accessible, all of which are free to use. Machine learning has grown into its own unique science, now covering the entirety of a language. This section's major objective is to find an effective machine learning technique by identifying a suitable machine learning approach for production applications.

Using machine learning in manufacturing has the following advantages: To better understand the constraints of Machine Learning (ML) in NP-complete settings, we will examine some of the benefits it has previously shown when optimizing for Smart Manufacturing Systems. Using advanced machine learning methods in a dynamic and complicated context, this paper examines how enormous datasets are linked together implicitly. Most technological and manufacturing issues have a lot of data, but little understanding. As a result, by utilizing ML, we are able to better grasp the issue at hand. There are several advantages listed in this section. These benefits are offered to provide you a basic understanding of machine learning. ML methodology can have a significant impact on the originality of the benefits. Most experts agree that ML enables for shorter cycle times, less garbage to be generated and more efficient resource use for NP-hard fabrication problems. ML software also offers significant quality improvement approaches that may be used in complex processes like electronics manufacture. Using machine learning algorithms has the advantage of being able to deal with complex problems and data. With any luck, it'll become even more valuable in the future (due to the rising availability of challenging data [such as social media accounts], and low production transparency). The point made here cannot be elaborated on further because it pertains to the great majority of the advantages and disadvantages associated with machine learning algorithms in general. SVM (Distributed Hierarchical Decision Tree) is one algorithm that can handle a greater number of dimensions than others. There have been numerous examples of machine learning algorithms used in manufacturing in the preceding section. A further benefit for machine learning is the ability to deal with large amounts of data. Modern machine learning approaches make use of technology like rapid miner to speed up the process and deliver new features. Classification performance is only little improved as a result of this, despite the fact that faster applications and more agreeable parameter adjustments are made possible. Explicit linkages between sets of data were addressed earlier. Data mining with machine learning algorithms has the added benefit of revealing previously undiscovered (implicit) information and uncovering previously unnoticed data connections. In particular, if an ML methodology is utilized (supervised, unsupervised, or reinforced learning), different methods have distinct data needs. Although significant results have been achieved in real-world industrial applications, it may be said that the ML algorithm can build real-world products. Expanding the capabilities of manufacturing systems is crucial since they are dynamic, diversified, and complicated. Because of the dynamic nature of the system and the algorithms' ability to learn and change on their own, this is only possible with ML algorithms. Some machine learning (ML) algorithms change far more quickly than older methods, and this is virtually always the case.

In the creation of ML models, patterns from earlier data sets may emerge, which could serve as a point of reference for predicting future system behavior. Having access to this additional information will help you make better judgments and improve the efficiency of the system as a whole. Identifying a pattern or regularity in a dataset is the ultimate goal of all of these distinct types of machine learning. The study used a variety of methodologies based on several variables to gauge production efficiency. See for yourself that this is an exploratory investigation. Tables like these should not be used to select a data collection because they show multiple perspectives on the same data. Data or pre-processing and parameter options determine how well an algorithm performs for a given issue. Before settling on the optimal approach, it's necessary to do a realistic test on various different algorithms to see which perform the best.

## Machine Learning has to Contend with Manufacturing Application Problems

Another major problem in the design of Application domains in manufacturing is the inability to collect the correct data. Availability, quality, and composition are all constraints we must deal with. The performance of the algorithm is directly related to the quality of the production data supplied. High-dimensional data, for example, may contain a substantial amount of duplicated and redundant information that interferes with the performance of machine learning. A wide range of machine learning algorithms are now accessible, but they all require historical data to work. The outcome is based on a variety of factors, including the method and settings employed. Simply put, when it comes to data gathering, most industrial research (or ML in general) may encounter a number of difficulties, especially if security issues or a lack of data collecting are present. Despite the fact that machine learning generally allows for greater data extraction and better results than traditional approaches, certain data problems can hinder the successful application of the technology. The second point emphasizes the increased importance of understanding the data in order to utilize machine learning techniques. Data is used to train machine learning algorithms rather than retrieving information as is the case with traditional methodologies. As soon as the data is accessible, it must be processed.

Ensure that you do this in accordance with your preferred method(s). The results of data preprocessing are highly influenced. When it comes to data normalization and filtering, all you need are a few standard tools to help. Checking for imbalances in the training samples is also required. With algorithm development, filling in these gaps can be a real challenge. Data value concerns have long plagued manufacturing; in particular, data values required to establish numerous features (e.g., product size, packing information, manufacturing method, and material) have been incomplete or missing for a long time. They are. Even though lacking data makes using ML algorithms more difficult, they can nevertheless be beneficial in certain situations. Effective induction methods are readily available to fill the void. There are distinct requirements for substituting missing values for each problem and consequently for the different machine-learning algorithms. We added a new value to fill in for a previously absent one in the original dataset. The objective is to minimize the prejudice and other side effects that drugs may produce. As long as this issue recurs, there are a plethora of remedies and resources to be found. Choosing a machine learning model and method to utilize is one of the most critical issues. Generic ML approaches, which are procedures that are universal in nature yet may be applied to a wide range of problems, were also pursued. Concerns and their different needs highlight the necessity for methods with varying strengths and shortcomings. More and more industrial and academic professionals are placing a high value on machine learning, which results in a proliferation of machine learning algorithms and modifications thereto (Smith et al., 2016; Thrimurthulu et al., 2004).

More and more people are using hybrid methods, also called multi-algorithm or bipartite techniques, since they expect better results than traditional single-purpose algorithms. Numerous studies have demonstrated the usefulness of machine learning techniques in a variety of different contexts. The findings of tests are not usually made public, which is another drawback. A comparison study of data becomes a subjective exercise since objectivity and impartiality are impossible to achieve. Many people believe in using machine learning (ML) to solve specific problems, yet no one appears to agree on the optimal ML strategy.

First, look at the available data and how it's described to choose whether to use a supervised, unattended, or RL technique. It is also necessary to investigate the general applicability of present methodologies to research topics. There must be a focus on the data structure, classifications, and overall amount of data available for training and assessment. Third, in order to find an acceptable method, earlier applications of algorithms on comparable situations should be explored. Similar issues are those that have criteria that are the same across multiple disciplines or fields. The interpretation of the results is also an issue. Keeping these aspects in mind when doing interpretation is crucial since it depends not only on the format or portrayal of results but also on the specifications for technique, values of parameters, expected results, and the data.

If the algorithm is changed, additional specific constraints may have a major impact on how the findings are interpreted. Excess fitting, bias, and variance immunity are only a few examples.

## CONCLUSIONS AND FUTURE PROSPECT

Machine learning (ML) is beneficial in the Additive Manufacturing process, according to research. Design, planning and control optimization of additively manufactured processes, quality control and enhancement of additively manufactured products can all benefit from machine learning. Additive manufacturing's benefits and drawbacks, as well as other issues impeding its use, must be addressed, as must the associated issues. Machine learning (ML) has been shown in research to be effective in improving AM design, process, and production. ML has been used in the design of AM parts to speed up tools, investigate new materials, identify property–structure correlations, and assist new designers in improving product quality. In the current state of affairs, there is only so much that can be done in terms of acceleration and material exploration. However, current implementations suffer from insufficient transitional zones and inaccuracies, which may cause some industries to consider it too low. When using the A.M. method, the majority of the research is devoted to optimizing process parameters. These are effective for improving production process parameters for one or more quality indicators. In a nutshell, ML is improving the prospects for AM adoption, and this trend is expected to continue.

In general, machine learning has benefited the prospects for increasing AM adoption and improving its value proposition. Having said that, the majority of machine learning applications for AM are not robust or trustworthy enough to be adopted in industry. As a result, research efforts should be directed toward further developing these tools for real-world application and publishing industry case studies to demonstrate their efficacy.

## REFERENCES

Boschetto, V., Giordano, V., & Veniali, F. (2013). Surface roughness prediction in fused deposition modelling by neural networks. *International Journal of Advanced Manufacturing Technology*, *67*(9-12), 2727–2742. doi:10.100700170-012-4687-x

Garg, A., & Tai, K. (2014). An Ensemble Approach of Machine Learning in Evaluation of Mechanical Property of the Rapid Prototyping Fabricated Prototype. *Applied Mechanics and Materials*, *575*, 493–496. doi:10.4028/www.scientific.net/AMM.575.493

Jaya Christiyan, K. G., Chandrasekhar, U., & Venkateswarlu, K. (2016). A study on the influence of process parameters on the Mechanical Properties of 3D printed ABS composite. *IOP Conf. Series: Materials Science and Engineering, 114*. 10.1088/1757-899X/114/1/012109

Keshavamurthy, Tambrallimath, Ugrasen, & Girish. (n.d.). Sustainable Product Development by Fused Deposition Modelling Process. *Fused Deposition Modeling Based 3D Printing*, 213-225.

Keshavamurthy, Vijay, & Saravanabavan. (n.d.). Development of Polymer Composites by Additive Manufacturing Process. In *Reference Module in Materials Science and Materials Engineering*. Doi: doi:10.1016/B978-0-12-803581-8.11885-5

Liu, C., Roberson, D., & Kong, Z. (2017). Textural Analysis-Based Online Closed-Loop Quality Control for Additive Manufacturing Processes. *2017 Industrial and Systems Engineering Conference*, Blacksburg, VA.

Ning, Cong, Qiu, Wei, & Wang. (2015). Additive Manufacturing of carbon fiber reinforced thermoplastic composites using fused deposition modeling. *Composites Part B, 80*, 369-378.

Phatak, A. M., & Pande, S. S. (2012). Optimum part orientation in rapid prototyping using genetic algorithm. *Journal of Manufacturing Systems, 31*(4), 395–402. doi:10.1016/j.jmsy.2012.07.001

Razvi, S. S., Feng, S., Narayanan, A., Lee, Y.-T. T., & Witherell, P. (2019). A Review of Machine Learning Applications in Additive Manufacturing. *Proceedings of the ASME 2019 International Design Engineering Technical Conferences and Computers and Information in Engineering Conference IDETC/CIE2019*. 10.1115/DETC2019-98415

Smith, J., Xiong, W., Yan, W., Lin, S., Cheng, P., Kafka, O. L., Wagner, G. J., Cao, J., & Liu, W. K. (2016). Linking Process, Structure, Property, and Performance for Metal-Based Additive Manufacturing: Computational Approaches with Experimental Support. *Computational Mechanics*, *57*(4), 583–610. doi:10.100700466-015-1240-4

Sood, A.K., Ohdar, R., & Mahapatra, S. (2013). Parametric appraisal of mechanical property of fused deposition modelling processed parts. *Materials & Design, 31*(1), 287-95.

Tambrallimath, Keshavamurthy, Patil, & Adarsha. (n.d.). Mechanical and Tribological Characteristics of Polymer Composites Developed by Fused Filament Fabrication. *Fused Deposition Modeling Based 3D Printing*, 151-166.

Tambrallimath, Keshavamurthy, Saravanbavan, Pradeepkumar, & Kumar. (2019). Synthesis and Characterization of Graphene Filled PC-ABS Filament for FDM Applications. Advances in Polymer Composites. Mechanics, Characterization and Applications AIP Conf. Proc., 2057.

Thrimurthulu, K., Pandey, P. M., & Reddy, N. V. (2004). Optimum part deposition orientation in fused deposition modeling. *International Journal of Machine Tools & Manufacture*, *44*(6), 585–594. doi:10.1016/j.ijmachtools.2003.12.004

Wang, C., Tan, X. P., Tor, S. B., & Lim, C. S. (2020). Machine learning in additive manufacturing: State-of-the-art and perspectives. *Additive Manufacturing*, *36*, 101538. doi:10.1016/j.addma.2020.101538

Wang, X., Jiang, M., Zhou, Z., Gou, J., & Hui, D. (2017a). 3D printing of polymer matrix composites: A review and prospective. *Composites. Part B, Engineering*, *110*, 442–458. doi:10.1016/j.compositesb.2016.11.034

Witherell, P. (2018). Emerging Datasets and Analytics Opportunities in Metals Additive Manufacturing. *Direct Digital Manufacturing Conference*, Berlin, Germany.

Wu, H., Wang, Y., & Yu, Z. (2015). In Situ Monitoring of FDM Machine Condition via Acoustic Emission. *International Journal of Advanced Manufacturing Technology*. Advance online publication. doi:10.100700170-015-7809-4

Wu, H., Yu, Z., & Wang, Y. (2017). Real-Time FDM Machine Condition Monitoring and Diagnosis Based on Acoustic Emission and Hidden Semi-Markov Model. *International Journal of Advanced Manufacturing Technology*, *90*(5–8), 2027–2036. doi:10.100700170-016-9548-6

Yia, Gläßnera, & Krenkela. (2019). *Energy simulation of the fused deposition modeling process using machine learning approach*. Academic Press.

Zhu, Z., Anwer, N., Huang, Q., & Mathieu, L. (2018). Machine learning in tolerancing for additive manufacturing. *CIRP Ann. Manuf. Technol.*, *67*(1), 157–160. doi:10.1016/j.cirp.2018.04.119

# Chapter 2
# Parts Design and Process Optimization

**Hany Hassanin**
*Canterbury Christ Church University, UK*

**Prveen Bidare**
*University of Birmingham, UK*

**Yahya Zweiri**
*Center for Autonomous Robotic Systems, Department of Aerospace Engineering, Khalifa University, Khali, UAE*

**Khamis Essa**
*University of Birmingham, UK*

## ABSTRACT

*Artificial intelligence and additive manufacturing are primary drivers of Industry 4.0, which is reshaping the manufacturing industry. Based on the progressive layer-by-layer principle, additive manufacturing allows for the manufacturing of mechanical parts with a high degree of complexity. In this chapter, a deep learning neural network (DLNN) is introduced to rationalize the effect of cellular structure design factors as well as process variables on physical and mechanical properties utilizing laser powder bed fusion. The models developed were validated and utilized to create process maps. For both design and process optimization, the trained deep learning neural network model showed the highest accuracy. Deep learning neural networks were found to be an effective technique for predicting material properties from limited data sets, as per the findings.*

DOI: 10.4018/978-1-7998-8516-0.ch002

## INTRODUCTION

Two essential pillars of Industry 4.0, which is transforming the manufacturing industry's paradigm, are additive manufacturing (AM) and deep learning (DL). Based on the incremental layer-by-layer principle, additive manufacturing allows for the manufacturing of mechanical parts with a high degree of complexity and flexibility (Davidson & Singamneni, 2016; Ge, Lin & Guo, 2018; Kayacan, Özsoy, Duman, Yilmaz & Kayacan, 2019; Scherillo et al., 2019). Laser powder bed fusion (L-PBF) has been widely used in a variety of industries, including biomedical (Brambilla, Okafor-Muo, Hassanin & ElShaer, 2021; Hany Hassanin, Modica, El-Sayed, Liu & Essa, 2016; Langford, Mohammed, Essa, Elshaer & Hassanin, 2021; Okafor-Muo, Hassanin, Kayyali & ElShaer, 2020), aerospace (Galatas, Hassanin, Zweiri & Seneviratne, 2018; Hany Hassanin, Abena, Elsayed & Essa, 2020; Hany Hassanin, Alkendi, Elsayed, Essa & Zweiri, 2020; Klippstein, Hassanin, Diaz De Cerio Sanchez, Zweiri & Seneviratne, 2018), and automotive (Schmitt, Mehta & Kim, 2020), since it can produce high-quality components from a variety of materials, including metals, ceramics, and polymers (El-Sayed, Hassanin & Essa, 2016; K. Essa et al., 2017; H. Hassanin & Jiang, 2010; Jiménez et al., 2021; Mohammed, Elshaer, Sareh, Elsayed & Hassanin, 2020). In this L-PBF process, a fast-moving laser beam is employed as an energy source to selectively melt the metal powder, resulting in dense metal components. L-PBF technology has the potential to revolutionize metal component manufacturing by making it more cost-effective, efficient and faster. Statistical tools such as the design of experiments (DOE) are commonly used to investigate and optimize the influence of AM process parameters. Although these methods proved to be effective, a common flaw is that the AM process parameters were believed to be static, even though AM is a dynamic process (Khamis Essa, Khan, Hassanin, Attallah & Reed, 2016; Sabouri et al., 2017). AM is characterized by scattered results due to repeated heating and cooling cycles, inter-layer interactions, change in the heat distribution within the build platform, and oxygen ($O_2$) level variability, even if it is within a defined range (Olson, 1997).

On the other side, metal cellular structures are lightweight-engineered high-performance materials with a unique mix of high load-bearing capacity, high energy absorption, and thermal and acoustic insulation properties. These characteristics made them suitable for high-performance products such as filters, catalytic converters, acoustic absorbers, heat exchangers, abradable seals, porous burners, biomedical implants, and oil sensors (Delgoshaei, Ariffin, Leman, Baharudin & Gomes, 2016; Guillame-Gentil et al., 2010; Sabouri et al., 2017; Tan, Tan, Chow, Tor & Yeong, 2017; L. Yang et al., 2015). Periodic and stochastic porous structures are two types of porous structures. Pores in periodic lattice structures are homogenous because they are made up of repeated unit cells, but pores in stochastic porous structures are randomly dispersed. Because of their intrinsic defects, the mechanical properties of periodic lattice structures generally outperform those of stochastic porous structures. The complexity and time required to create periodic lattice structures using traditional production technologies such as casting and machining are very high and that prevents their widespread application (Al-Ketan, Soliman, AlQubaisi & Abu Al-Rub, 2018; Felzmann et al., 2012).

There are two numerical techniques that have been effectively utilized to forecast the behaviour of produced parts are finite element modelling and computational fluid dynamics. Belhocine et al. (Belhocine & Afzal, 2020) created a numerical simulation of structural analysis as well as a transient thermal coupled thermo-structural approach for brake disc rotors made of various materials. Due to the lack of research studies that address the aforementioned concerns, the research has shifted to use different methods. Deep learning (DL) has evolved in recent years, opening up new avenues for AM research (Yuan et al., 2018). However, because DL works best with large data, the small data nature of AM trials is a hindrance to its use in the analytical modelling of AM processes (Weichert et al., 2019).

Artificial neural networks are mathematical algorithms based on animal brain neural networks. One input layer, one hidden layer, and one output layer make up a shallow neural network (SNN). Deep learning neural networks (DLNNs) are neural networks with more than two layers (J. Yang et al., 2020). When modelling complicated situations, DLNNs are more efficient than SNNs because they use nonlinear activation functions at several levels (Bengio, Lamblin, Popovici & Larochelle, 2006). Local minima can be overcome using the greedy layer-wise pre-training method (Geoffrey E. Hinton, Osindero & Teh, 2006; G. E. Hinton & Salakhutdinov, 2006). Greedy layer-wise pre-training sets a neural network's weights to values in the vicinity of a local minimum. As a result, it facilitates the optimization process and leads to improved model generalization. In a pre-training method, a stacked auto-encoder (SAE) architecture was utilized as an alternative to the Boltzmann machine (Azzam, Taha, Huang & Zweiri, 2020).

As a result, the processing and characterization of Ti-6Al-2Sn-4Zr-6Mo alloy using L-PBF is the focus of this chapter. The research looks at how the L-PBF process parameters affect the characteristics of Ti-6Al-2Sn-4Zr-6Mo alloy. In addition, the study created various deep learning models for process optimization based on the measured data of the alloy processed in L-PBF, to predict the properties of the produced alloy. Laser power, scan speed, island size, and hatch spacing are among the process parameters investigated in this study. Deep learning was used to study and model the changes in porosity, hardness, and microstructure at various L-PBF processing settings. The machine-learning algorithms will also be used in this chapter to predict the properties of additively produced titanium alloy cellular structures and, as a result, optimize their topology. The size and orientation of the cells were taken into account as input parameters. As output data, the ultimate compressive strength, Young' modulus, and specific strength were investigated. In order to understand the influence of process and design parameters and to enable accurate prediction of properties when processed by L-PBF, shallow neural network supervised training, deep neural network supervised training, and deep learning neural network unsupervised greedy layer-wise pre-training approaches were used.

## MATERIALS AND METHODS

Figure 1 illustrates the flow diagram for processing Ti-6Al-2Sn-4Zr-6Mo alloy and the cellular structure.

### Ti-6Al-2Sn-4Zr-6Mo Manufacturing

In this process, a TLS, Bitterfeld, Germany, supplied Ti-6Al-2Sn-4Zr-6Mo powder was used, which was sieved in the 20–50 m range for this study. In order to prepare 10 mm x 10 mm cuboid samples for this study, a laser powder bed fusion system (M2 Concept Laser, Lichtenfels, Germany) with a Nd:YAG laser with a power of up to 200 W and a laser speed of up to 4000 mm/s was employed. Figure 2a shows a schematic design of the L-PBF system. All of the cuboid samples were produced on a titanium build substrate, in an argon chamber Oxygen ($O_2$) level less than 100 ppm. Figure 2b shows the island scanning method, which divides the laser-scanned portion into squares known as islands. In order to ensure repeatability, three samples were built for each run in all tests.

The input parameters employed in the matrix preparation were laser power of (100–200 W), laser speed of (800–1800 mm/s), hatch spacing constant h1 of (0.2–0.8), and island size of (2–8 mm). One of the essential terms utilized in L-PBF operations is volumetric energy density (E). It's an empirical parameter that represents the effect of L-PBF laser parameters on the properties of the samples. Eq.1 shows the equation for the volumetric energy density.

$$E = \frac{P}{v.h.b} \tag{1}$$

where 'P' is the laser power in watts (W), 'v' is the laser scan speed in mm/s, 'h' is the hatch spacing in mm, and 'b' is the thickness of the powder layer in mm. Table 1 shows the matrix parameters and levels that were generated.

*Figure 1. The flow diagram of the processing and modelling procedure.*

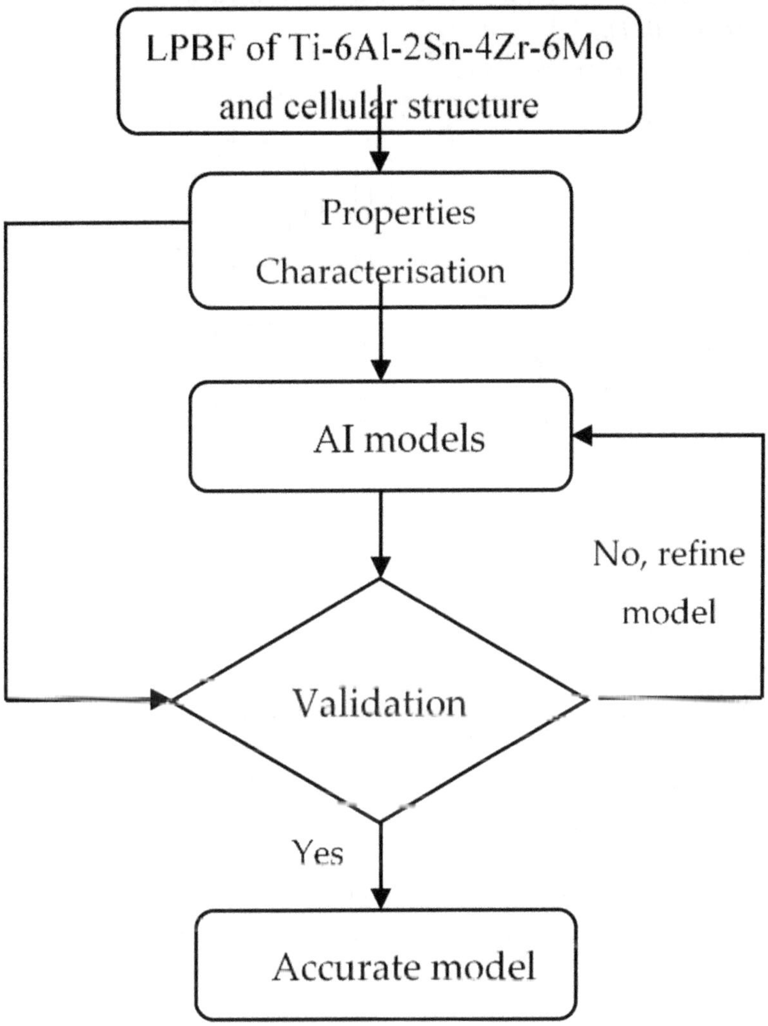


Hot isostatic pressing (HIP) was performed in an EPSI HIP vessel at the University of Birmingham, which has a maximum temperature capacity of 1450 °C and a maximum pressure capacity of 200×10$^6$ Pa. A water-cooled vessel with molybdenum heating components and a compressed Argon gas system makes up the HIP unit. In this experiment, the HIP cycle was 800 °C/103 MPa/4 h, followed by furnace cooling.

In order to produce cross-sections of building layers, the Ti-6Al-2Sn-4Zr-6Mo samples were cut vertically across the X-Z plane into two parts. The standard grinding and polishing processes were used to polish metallographic samples. The polished

*Figure 2. (a) A schematic illustration of the laser powder bed fusion system; (b) contours for island scanning and hatch spacing, reused with permission (Hany Hassanin et al., 2021).*

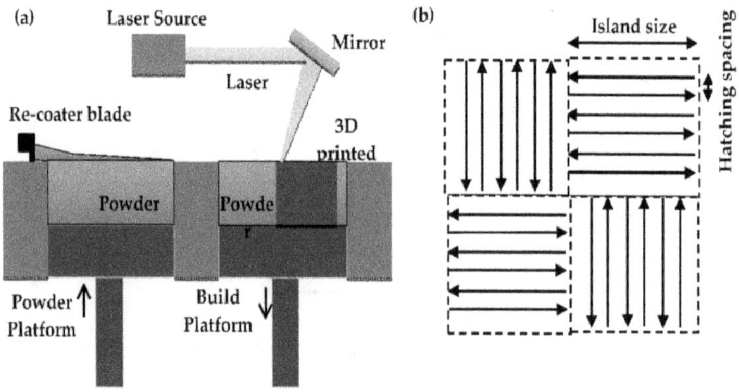


cross-sections were characterized for porosity using an optical microscope (OM) Zeiss Axioskop and a Hitachi TM300 (Hitachi, Japan) desktop electron scanning microscopy (Peine, Germany) (SEM). Using ImageJ (an image editor), more than 80 images were acquired and stitched together to create the majority of the cross-section. The software was used to calculate the porosity's fractional area. Vickers micro-hardness tests were carried out using a 30 kg indenter load on an INDENTEC hardness testing equipment (Brierley, UK). The phase development between the as SLMed and HIPed samples was investigated using X-ray diffraction (XRD) utilizing an Inel EQUINOX 3000 (Waltham, Massachusetts, United States) with a Cu-fiber laser of 1.54.

*Table 1. The designed process parameters.*

| Process Parameter | Units | Levels | | | | |
|---|---|---|---|---|---|---|
| | | −2 | −1 | 0 | 1 | 2 |
| **Laser power** | W | 100 | 125 | 150 | 175 | 200 |
| Scan speed | mm/s | 800 | 1050 | 1300 | 1550 | 1800 |
| Hatch spacing | ($h_l$) | 0.2 | 0.35 | 0.5 | 0.65 | 0.8 |
| Island size | mm | 2.0 | 3.5 | 5.0 | 6.5 | 8.0 |

## Cellular Structure Fabrication and Characterisation

In order to achieve particular features, controlled density is demonstrated in periodic cellular structures with predetermined shape (Zhang, Leary, Tang, Song & Qian, 2018). Periodic lattice structures have design characteristics such as density, pore size, and feature size. Periodic lattice structures include the Gyroid, Diamond, and Neovius (E. Yang et al., 2019). They differ in terms of strut shape, orientation, thickness, and nodal connection. The diamond shape lattice structure is the most advantageous of these sorts and has thus been used in a variety of applications (K. Essa et al., 2017; Tamburrino, Graziosi & Bordegoni, 2018).

The diamond-shaped lattice structure comprises four struts that are nodally connected to another four struts, allowing for tremendous flexibility in modifying the volume fraction without the usage of support structures. The nodally-connected diamond lattice structure and its unit cell are shown in Figure 3. The strut length (L), strut diameter (D), and strut orientation angle theta (θ) are the studied design parameters of this structure employed in this work. Table 1 displays the design parameters and levels that were developed. The levels values in Table 1 were chosen based on a mix of sample manufacturability and geometrical restrictions to achieve compression samples in accordance with ASTM. In this chapter, a design of experiment (DoE) strategy with a response surface is employed as a reference approach.

The Ti-64 material was used in the production of the lattice structures. The material is lightweight and has excellent mechanical properties, making it ideal for aerospace and biomedical applications. The proposed lattice matrix was created using laser powder bed fusion (L-PBF). Ti6Al4V powder with a particle size range of 25–50 m was supplied by (TLS Technik GmbH, Germany). A Concept Laser M2 L-PBF machine is employed, which has a 1075 nm Nd:YAG laser with a beam spot of 50 m. The samples were produced utilizing Ti6Al4V OEM provided standard process parameters, including a laser power of 200 W, a scan speed of 1200 mm/s, and a layer thickness of 20 microns for compression testing. The samples were built on a Ti-6Al-4V base plate and under Argon controlled atmosphere down to $O_2$<100 ppm. Before compression testing, the samples were sonicated for 5 minutes in an acetone bath to release trapped particles and contaminants. The sample's density was measured using the Archimedes method and a Mettler-Toledo densitometer. The manufactured samples were compressed at room temperature using the Zwick/Roell method. Specimens containing 558 unit cells were crushed along the Z-axis (build direction). The tests were performed at a contact moving speed of 0.1 mm/min. The compressive maximum strengths are expressed using the nominal engineering strength (the maximum force divided by the cross-sectional area of the produced samples). This method reflects the samples' load-bearing capacity. The specific strength of the samples was calculated by dividing their maximum compressive

*Figure 3. The Nodally connected diamond structures employed in this study and the relevant design parameters, reused with permission (Hany Hassanin, Alkendi, et al., 2020).*

(a)

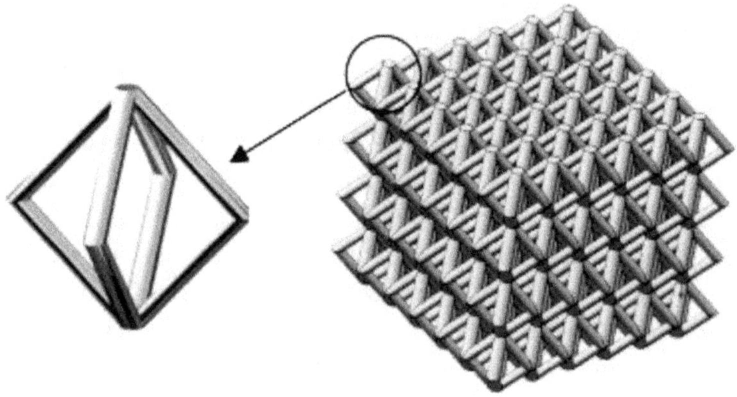

(b)

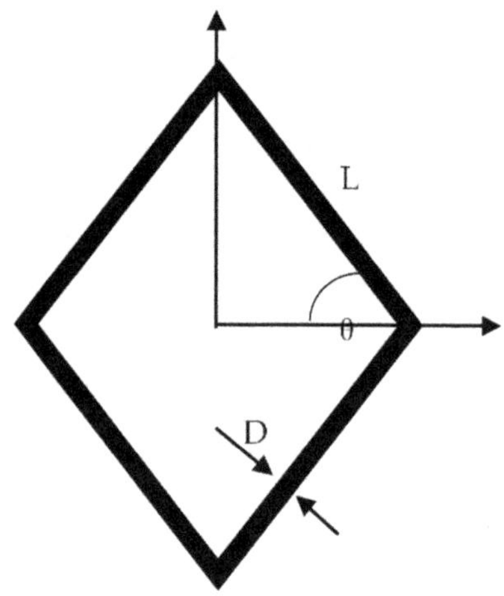


*Table 2. Design parameters and their corresponding levels*

| Parameter | Units | Levels | | | | |
|---|---|---|---|---|---|---|
| | | −2 | −1 | 0 | 1 | alpha |
| Strut Length (D) | mm | 0.2 | 0.36 | 0.6 | 0.83 | 1 |
| Strut diameter (L) | mm | 1 | 1.2 | 2.25 | 3.29 | 4 |
| Strut Orientation Angle (θ) | degree | 30 | 36.1 | 45 | 53.9 | 60 |

strength by their measured density. Furthermore, the samples' Young's modulus was calculated by dividing the nominal stress by the strain.

## Deep Learning

In this section of the chapter, the structure of the proposed deep learning neural network (DLNN) in terms of depth, hidden layer size, and activation functions is provided. The strategies for pre-training and fine-tuning are then explained. Ultimately, the backpropagation algorithm is discussed. Figure 1 shows a shallow neural network (SNN) that is given and utilized as a comparison to examine the performance of the developed DLNN.

Figure 2 shows the structure of the developed deep learning neural network (DLNN), which outperformed several other evaluated structures with various hidden layers, activation functions, and neuron sizes. An input layer, three hidden layers, and an output layer make up the system. The length, diameter, and orientation angle of the struts are all inputs to the input layer. The sigmoid function activates the output layer, which consists of three nodes. The Young's modulus, ultimate strength and specific strength are the network's outputs. Each of the three hidden levels has 50 neurons and is the same size. The authors of [19] recommended using hidden layers of the same size since it is more convenient when using the pre-training technique that will be discussed next. Swish was selected as the activation function for the first hidden layer, followed by a rectified linear unit (ReLU) for the second and third hidden layers, as shown in equations (2) and (3), respectively.

$$f_1(x) = x.sigmoid(x) = \frac{x}{1+e^{-x}} \tag{2}$$

$$f_2(x) = max(0,x) \tag{3}$$

Since the assigned weights have such a large impact on a neural network's performance [24], it's critical to ensure that the initial weights are in the vicinity of a good estimate so that the gradient descent method used during training converges to the local minimum area without the vanishing gradient issue (Bengio et al., 2006; G. E. Hinton & Salakhutdinov, 2006). To this purpose, a pre-training method is developed in the proposed approach to initialize the network's weights. The chosen pre-training technique is the developed unsupervised greedy layer-wise and its process is shown in Figure 4. The proposed DLNN is pre-trained in four steps, with the non-input layers of the DLNN being trained in sequence, starting with the first hidden layer and ending with the output layer. Each layer is trained separately using a Shallow Neural Network (SNN), also known as an auto-encoder, with appropriate input and output layers.

Although pre-training improves the initialization of the DLNN's weights, it is inefficient, especially for the weights in the first hidden layers (Geoffrey E. Hinton et al., 2006). As a result, using the backpropagation technique, global network fine-tuning is performed to replace random weights coming from unsupervised learning in the pre-training process with more deterministic ones (Alzahrani, Choi & Rosen, 2015). When compared to optimizing random beginning weights, the optimization process in this situation is significantly simpler and achieves superior generalization (Yan, Hao, Hussein & Raymont, 2012). Backpropagation (Gu, Chen, Richmond & Buehler, 2018) is a supervised learning strategy for neural networks in which the discrepancies between the DLNN's predictions and the related target outputs are utilized to adjust the network internal weights, refining the network parameters for optimal performance. The backpropagation algorithm is described in the following sections. The network's input is first forward propagated through the DLNN, where each layer computes its output as a function of the output of the layer before it. Then, the error between the predicted output and the target is determined using one or more of the cost functions (i.e., Mean Squared Error (MSE), Mean Absolute Error (MAE), or Cross-Entropy (CE) loss functions) and propagated backward to the network layers during the backdrop process. The updated weights are obtained by multiplying the derivative of the activation function with the estimated error using the gradient decent (GD) (between network predictions and targets). The minimum of a least square cost function would be obtained using the gradient decent approach and the backpropagation algorithm.

## RESULTS AND DISCUSSIONS

### Optimisation of Deep Learning (DL) Models

An autonomous search for the best optimal deep learning model was performed by iterating the initial number of layers, random seeds, number of neurons, and the activation functions. A variety of DNN structures with three, four, and five layers were trained and evaluated. Figure 4 depicts the structure with the lowest mean absolute error. Backpropagation of the resulting structure was compared to unsupervised greedy layer-wise pre-training. After that, the best model was compared to DNN and SNN. The DNN model was chosen to be identical to the optimal DLNN. The mean percentage error (MPE) for the tested techniques is shown in Table 3. In

*Figure 4. The shallow neural network supervised training, reused with permission (Hany Hassanin, Alkendi, et al., 2020)*

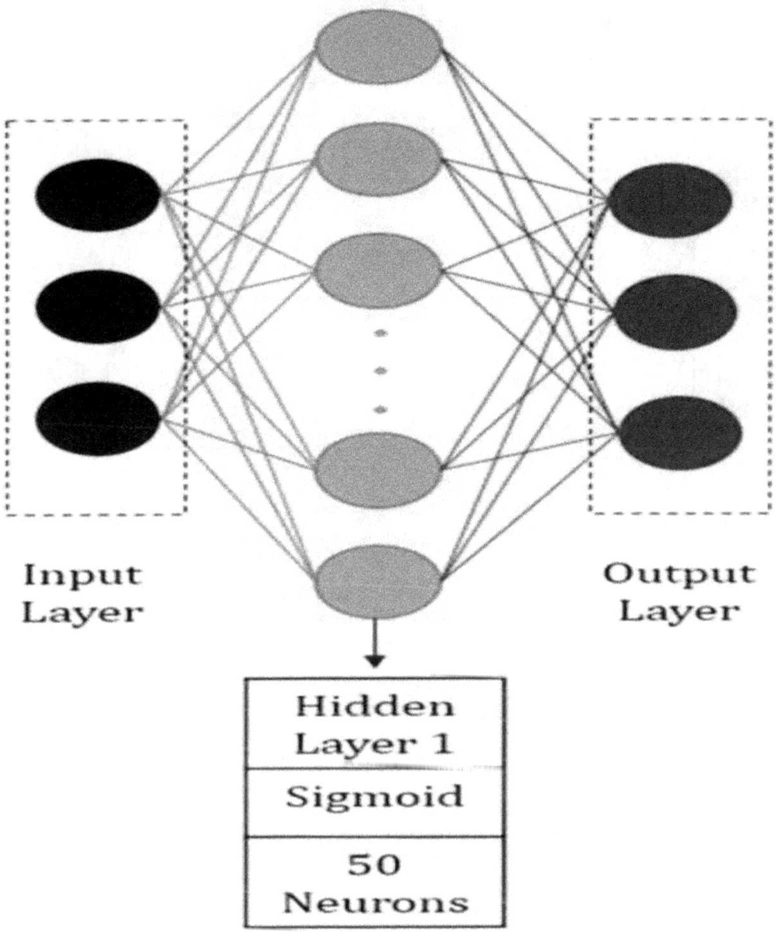


comparison to other models, the deep learning neural network (DLNN) structure in Figure 3 produced the lowest error.

## Validation

The mean percentage error values in Table 3 indicate model predictions based on only 90% of the experimental data. Figure 7-10 depicts a validation comparison between the model developed using a deep learning neural network unsupervised

*Figure 5. The deep neural network supervised training, reused with permission (Hany Hassanin, Alkendi, et al., 2020)*

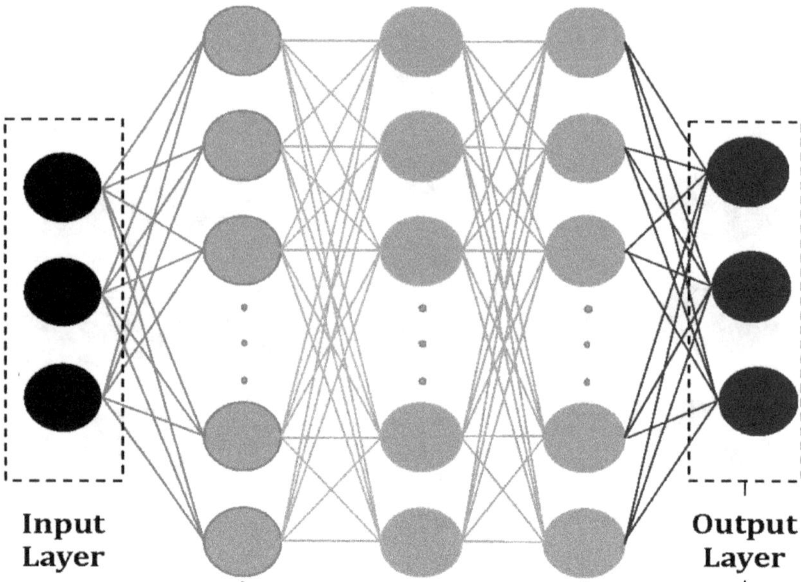


| Hidden Layer 1 | Hidden Layer 2 | Hidden Layer 3 | Output Layer |
|---|---|---|---|
| Swish | ReLU | ReLU | Sigmoid |
| 50 Neurons | 50 Neurons | 50 Neurons | 3 Neurons |

greedy layer-wise pre-training method and 100% of the measured data for both porosity and hardness. The red and blue lines in the figures show a strong agreement between the measured data and the DLNN model.

## Microstructural Analysis

The optimized DLNN model was used to create a contour map of the porosity alloy in relation to the process parameters, as shown in Figure 12. The contour distribution illustrates the areas with the least and most porosity, denoted as AL and AH. In the microstructure of the samples, defects such as lack of fusion porosity and keyholes

*Figure 6. The deep learning neural network unsupervised greedy layer-wise pre-training approach, reused with permission (Hany Hassanin, Alkendi, et al., 2020)*

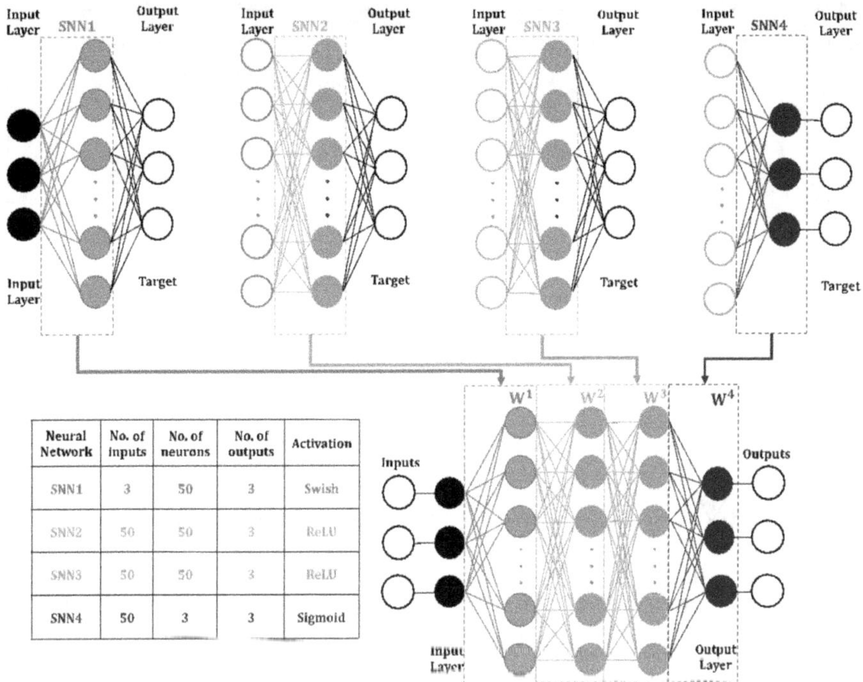


were revealed. When the energy density is low, melt pools do not overlap, leaving unmelted powder and forming fusion defects, whereas when the energy density is high, melt pools become unstable and deepen, forming keyhole pores. It was also found that when the energy density is high, the island size has an effect on the porosity of the sample, despite the fact that its original purpose was to disperse

*Table 3. Mean percentage error comparison for tested approaches.*

| Approach | Porosity | Hardness | Strength | Specific Strength | Youngs' Modulus |
|---|---|---|---|---|---|
| DLNN | 3% | 0.3% | 0.8% | 6.2% | 5% |
| DNN | 18% | 12% | 5% | 20% | 18% |
| SNN | 46% | 36% | 44% | 83% | 122% |

*Figure 7. Porosity was measured and compared to the DLNN model's output, reused with permission (Hany Hassanin et al., 2021).*

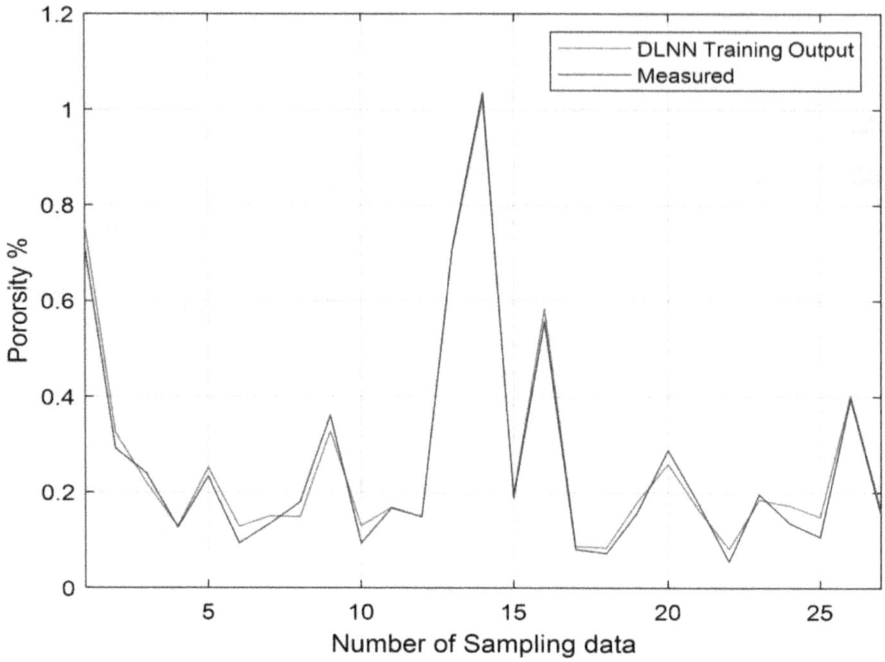


heat energy equally across the component cross-section and therefore minimize the produced thermal stresses.

## Hardness

Figure 13 presents the projected contour map of hardness versus process parameters using the optimum DLNN model. The effect of the island size parameter on hardness was found to be significant, similar to the effect of porosity.

The contour map of compressive ultimate strength and elastic modulus as a function of strut diameter and length is shown in Figure 14. Furthermore, the figure compares the stress-strain graphs of lattice samples produced with varying strut diameter and strut length (X1-X4). X1 and X2 are lattice samples produced with various strut diameters but the same strut length and orientation angle, whereas X3 and X4 are lattice samples produced with various strut lengths but the same strut diameter and orientation angle. Particularly, X1 and X2 are lattice structures with 2.25mm strut lengths and 0.2mm and 0.6mm strut diameters, respectively. X3 and X4, on the other hand, are lattice structures with strut diameters of 0.84 mm and strut lengths of 1.21mm and 3.29mm, respectively. In general, average compressive

*Figure 8. The measured hardness as compared to the DLNN model's output, reused with permission (Hany Hassanin et al., 2021).*

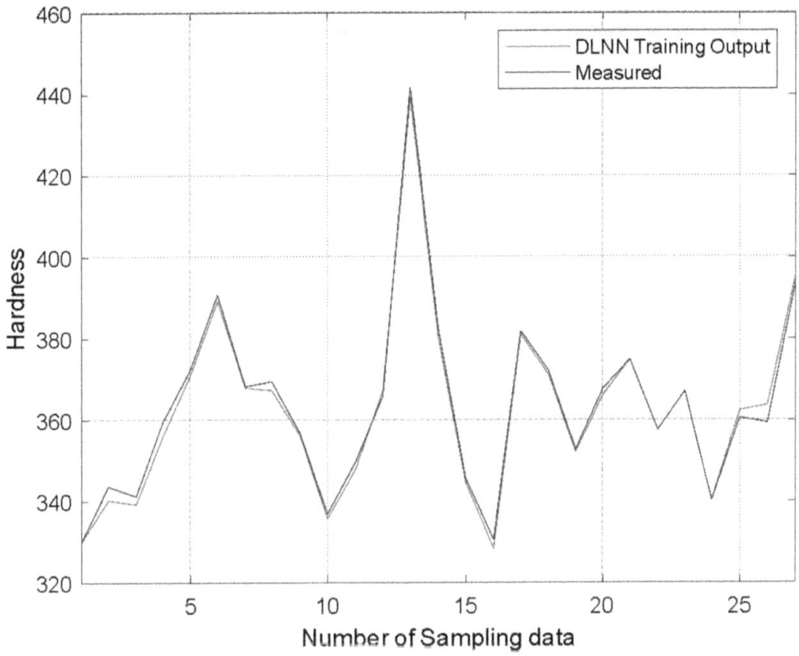


strength improves as strut diameter increases and strut length decreases. Likewise, the elastic modulus increases as the strut diameter increases and the strut length decreases.

The contour map of compressive ultimate strength and elastic modulus as a function of strut diameter and orientation angle is shown in Figure 15. Furthermore, the figure compares the stress-strain graphs of lattice samples produced from different strut angles (X5-X6). The graph indicates that as the strut angle increases, the compressive strength decreases. The elastic modulus, on the other hand, drops as the strut diameter and length increase. Figure 16 depicts the projected contours distribution of specific strength with respect to the diameter, length, and angle of the lattice struts. The graph exhibits a similar pattern to the prediction of ultimate strength. The specific strength of the samples improves as the diameter increases, whereas it falls as the strut length and angle increase.

*Figure 9. Maximum strength as compared to the DoE/DLNN/DNN/SNN models' output.*

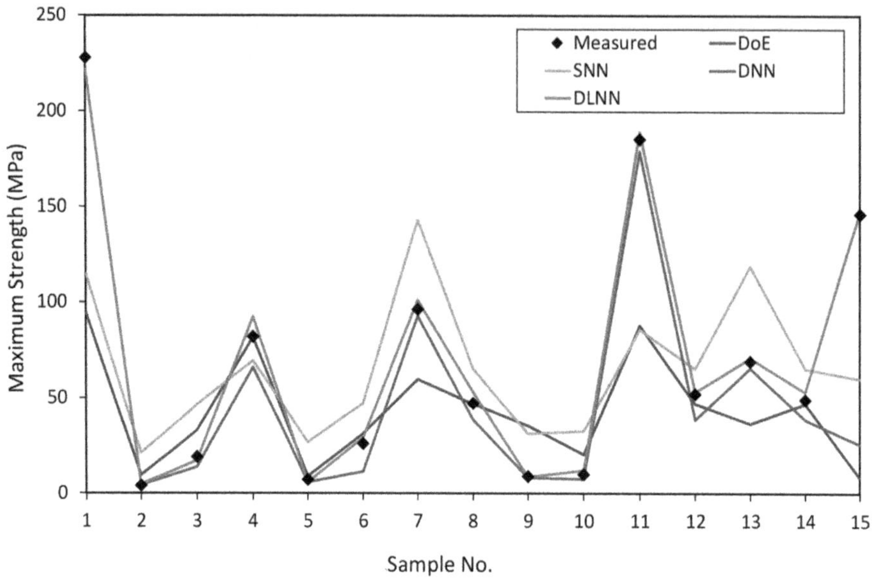


## CONCLUSION

In this chapter, the influence of L-PBF parameters like laser power, scan speed, island size, and hatch spacing, as well as cellular structure design parameters like strut length, strut diameter, and strut orientation angle, on the performance of processed items were studied. The DLNN model was found to accurately predict the properties of manufactured materials, designs, and distribution models, resulting in stronger and more controllable structures. The discussed results in this study lead to the following conclusions:

- The trained DLNN model demonstrated the best performance in comparison with the deep neural network (DNN), shallow neural network (SNN) and design of experiment (DoE).
- The porosity increased as the volumetric energy density decreased from 77 J/mm3 or increased further from 113 J/mm3. The Hardness of the L-PBF specimens increased as volumetric energy density increased.
- As the strut diameter increases and the strut length decreases, the compressive of the cellular structures increases. Furthermore, when the strut diameter and length increase, the elastic modulus increases.

*Figure 10. Specific strength compared to the output of the DLNN/DNN/DoE/SNN models, reused with permission (Hany Hassanin, Alkendi, et al., 2020).*

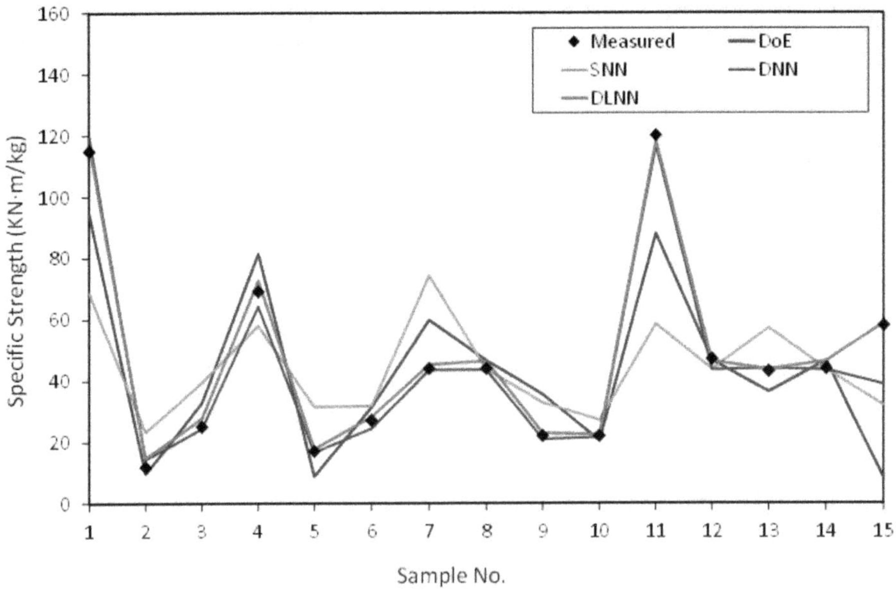


- The cellular structure's specific strength increases as the diameter increases, whereas it decreases as the strut length and angle increase.

The trained DLNN model predicted the occurrence of porosity level with an accuracy of 3%, 0.3%, 0.8%, 6.2%, and 5% for ultimate strength, elastic modulus, specific strength, Porosity, and Hardness. These findings indicate that the developed DLNN models are effective tools for learning measurement from limited measurement data and making high-fidelity performance predictions. The approach presented in this research can also be used for other types of cellular structures.

*Figure 11. The output of the DLNN/DNN/DoE/SNN models is compared to the Young's modulus, reused with permission (Hany Hassanin, Alkendi, et al., 2020).*

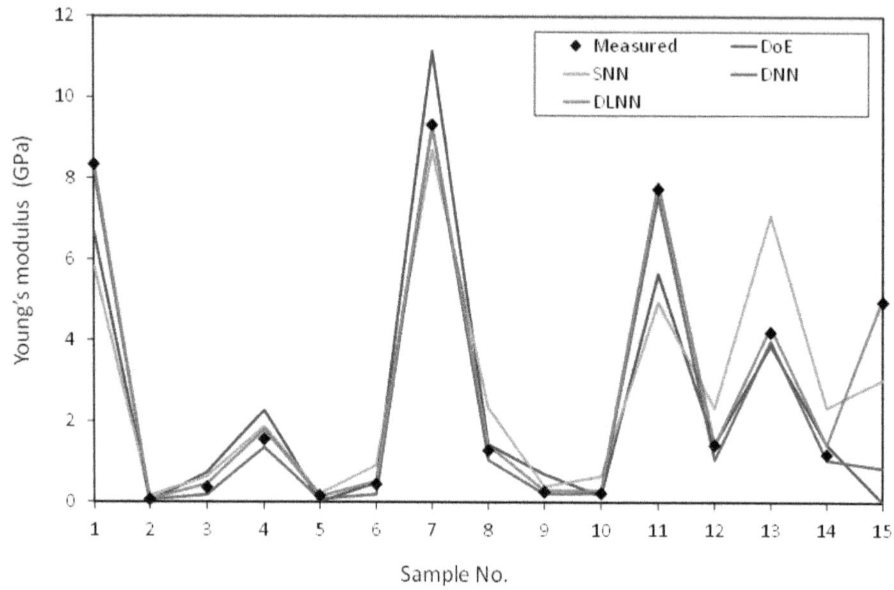


*Figure 12. The predicted porosity contours distribution of Ti-6Al-2Sn-4Zr-6Mo alloy with regards to process parameters and porosity defects (a) influence of power and scan speed, (b) effect of energy density and island size, reused with permission (Hany Hassanin et al., 2021).*

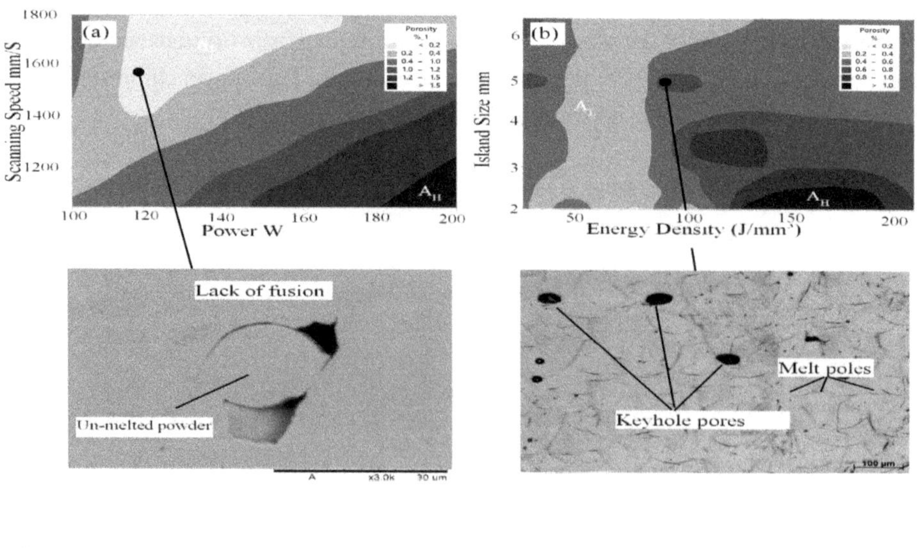

*Figure 13. Vickers microhardness of Ti-6Al-2Sn-4Zr-6Mo alloy predicted contours, (a) effect of power and scan speed, (b) effect of energy density and island size,, reused with permission (Hany Hassanin et al., 2021).*

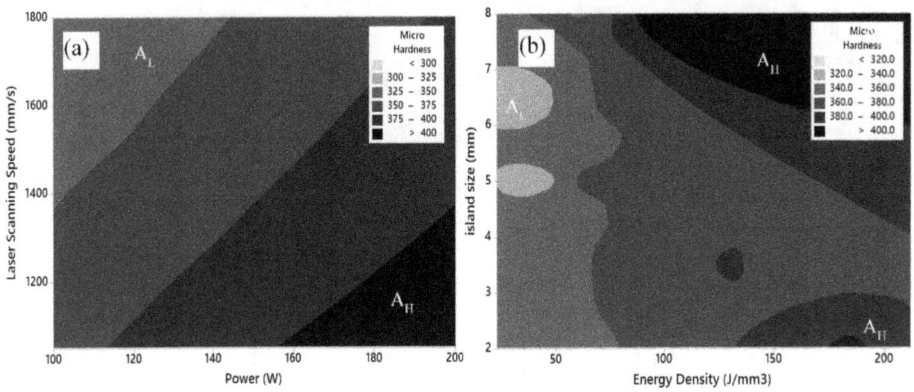


*Figure 14. (a-b) The predicted contours distribution of the ultimate compressive strength and elastic modulus with regards to the lattice struts diameter and length (c-d) stress-strain diagrams of samples X1-X4, reused with permission (Hany Hassanin, Alkendi, et al., 2020).*

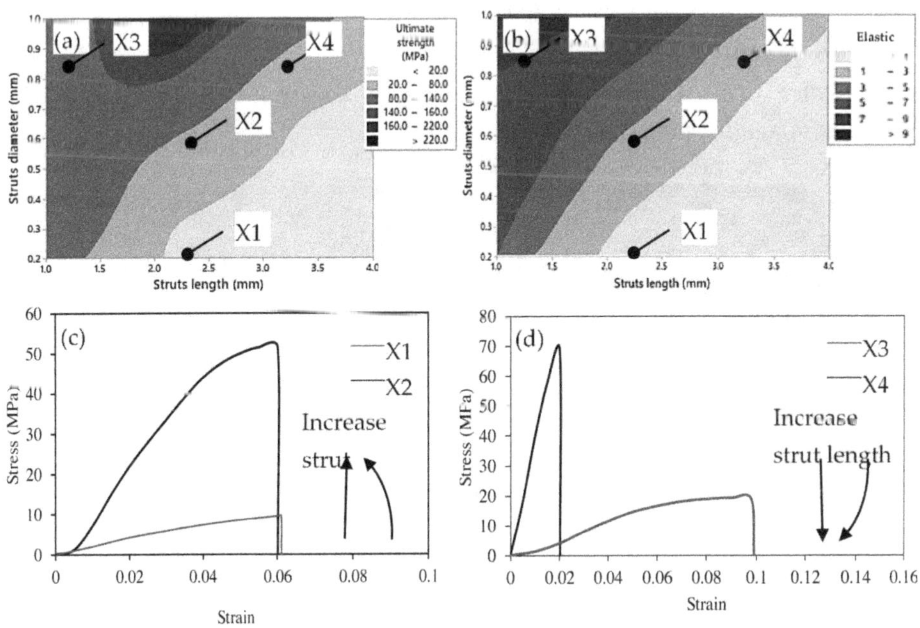

*Figure 15. (a-b) The predicted contours distribution of the ultimate compressive strength and elastic modulus with respect to the strut angle, (c) stress-strain diagrams of samples X5-X6, reused with permission (Hany Hassanin, Alkendi, et al., 2020).*

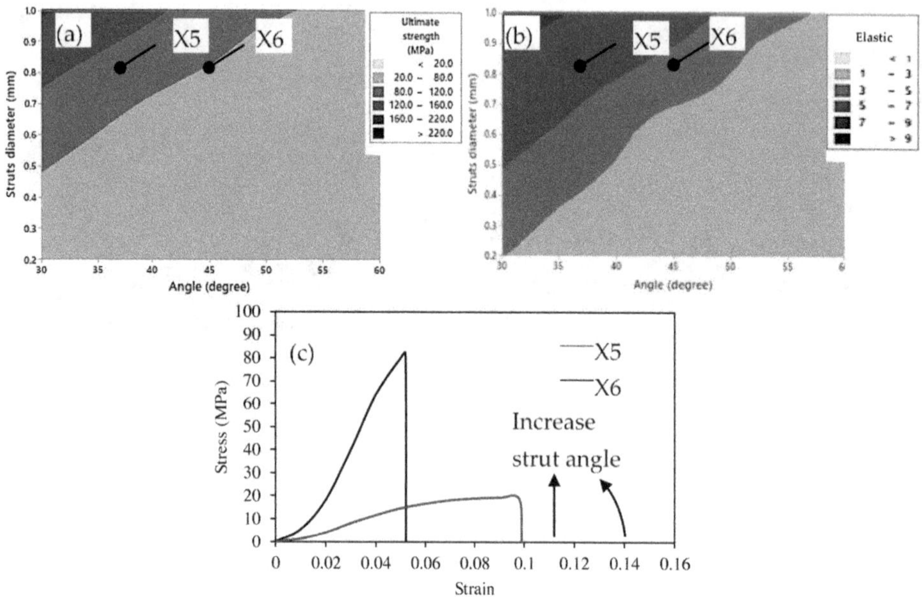


*Figure 16. The predicted contours distribution of the specific strength with regards to the lattice struts diameter, length, and angle, reused with permission (Hany Hassanin, Alkendi, et al., 2020).*

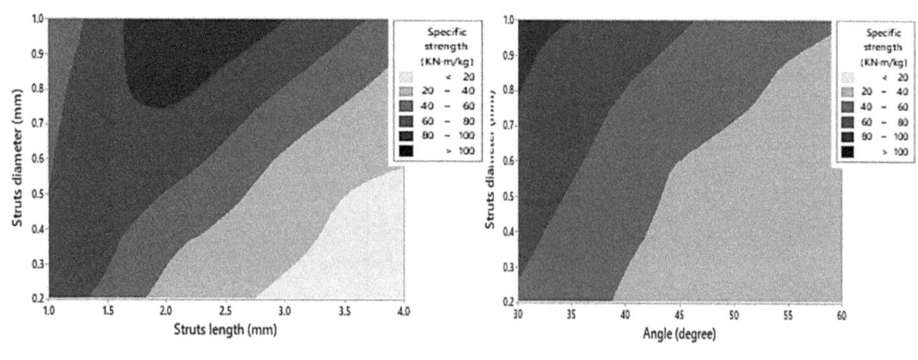

## REFERENCES

Al-Ketan, O., Soliman, A., AlQubaisi, A. M., & Abu Al-Rub, R. K. (2018). Nature-Inspired Lightweight Cellular Co-Continuous Composites with Architected Periodic Gyroidal Structures. *Advanced Engineering Materials, 20*(2), 1700549. Advance online publication. doi:10.1002/adem.201700549

Alzahrani, M., Choi, S. K., & Rosen, D. W. (2015). Design of truss-like cellular structures using relative density mapping method. *Materials & Design, 85*, 349–360. doi:10.1016/j.matdes.2015.06.180

Azzam, R., Taha, T., Huang, S., & Zweiri, Y. (2020). A deep learning framework for robust semantic SLAM. *2020 Advances in Science and Engineering Technology International Conferences, ASET 2020*. 10.1109/ASET48392.2020.9118181

Belhocine, A., & Afzal, A. (2020). Computational finite element analysis of brake disc rotors employing different materials. *Australian Journal of Mechanical Engineering*, 1–14. doi:10.1080/14484846.2020.1733175

Bengio, Y., Lamblin, P., Popovici, D., & Larochelle, H. (2006). Greedy layer-wise training of deep networks. *Proceedings of the 19th International Conference on Neural Information Processing Systems*.

Brambilla, C. R. M., Okafor-Muo, O. L., Hassanin, H., & ElShaer, A. (2021). 3DP Printing of Oral Solid Formulations: A Systematic Review. *Pharmaceutics, 13*(3), 358. https://www.mdpi.com/1999-4923/13/3/358

Davidson, K., & Singamneni, S. (2016). Selective Laser Melting of Duplex Stainless Steel Powders: An Investigation. *Materials and Manufacturing Processes, 31*(12), 1543–1555. doi:10.1080/10426914.2015.1090605

Delgoshaei, A., Ariffin, M. K. A. M., Leman, Z., Baharudin, B. T. H. T. B., & Gomes, C. (2016). Review of evolution of cellular manufacturing system's approaches: Material transferring models. *International Journal of Precision Engineering and Manufacturing, 17*(1), 131–149. doi:10.100712541-016-0017-9

El-Sayed, M. A., Hassanin, H., & Essa, K. (2016). Effect of casting practice on the reliability of Al cast alloys. *International Journal of Cast Metals Research, 29*(6), 350–354. doi:10.1080/13640461.2016.1145966

Essa, K., Hassanin, H., Attallah, M. M., Adkins, N. J., Musker, A. J., Roberts, G. T., Tenev, N., & Smith, M. (2017). Development and testing of an additively manufactured monolithic catalyst bed for HTP thruster applications. *Applied Catalysis A, General, 542*, 125–135. doi:10.1016/j.apcata.2017.05.019

Essa, K., Khan, R., Hassanin, H., Attallah, M. M., & Reed, R. (2016). An iterative approach of hot isostatic pressing tooling design for net-shape IN718 superalloy parts. *International Journal of Advanced Manufacturing Technology*, *83*(9), 1835–1845. doi:10.100700170-015-7603-3

Felzmann, R., Gruber, S., Mitteramskogler, G., Tesavibul, P., Boccaccini, A. R., Liska, R., & Stampfl, J. (2012). Lithography-based additive manufacturing of cellular ceramic structures. *Advanced Engineering Materials*, *14*(12), 1052–1058. doi:10.1002/adem.201200010

Galatas, A., Hassanin, H., Zweiri, Y., & Seneviratne, L. (2018). Additive Manufactured Sandwich Composite/ABS Parts for Unmanned Aerial Vehicle Applications. *Polymers, 10*(11), 1262. https://www.mdpi.com/2073-4360/10/11/1262

Ge, W., Lin, F., & Guo, C. (2018). Effect of energy input on microstructure and mechanical properties in EBSM Ti6Al4V. *Materials and Manufacturing Processes*, *33*(15), 1708–1713. doi:10.1080/10426914.2015.1048463

Gu, G. X., Chen, C.-T., Richmond, D. J., & Buehler, M. J. (2018). Bioinspired hierarchical composite design using machine learning: Simulation, additive manufacturing, and experiment. *Materials Horizons*, *5*(5), 939–945. doi:10.1039/C8MH00653A

Guillame-Gentil, O., Semenov, O., Roca, A. S., Groth, T., Zahn, R., Vörös, J., & Zenobi-Wong, M. (2010). Engineering the extracellular environment: Strategies for building 2D and 3D cellular structures. *Advanced Materials*, *22*(48), 5443–5462. doi:10.1002/adma.201001747 PMID:20842659

Hassanin, H., Abena, A., Elsayed, M. A., & Essa, K. (2020). 4D Printing of NiTi Auxetic Structure with Improved Ballistic Performance. *Micromachines, 11*(8), 745. https://www.mdpi.com/2072-666X/11/8/745

Hassanin, H., Alkendi, Y., Elsayed, M., Essa, K., & Zweiri, Y. (2020). Controlling the Properties of Additively Manufactured Cellular Structures Using Machine Learning Approaches. *Advanced Engineering Materials*, *22*(3), 1901338. doi:10.1002/adem.201901338

Hassanin, H., & Jiang, K. (2010). Optimized process for the fabrication of zirconia micro parts. *Microelectronic Engineering*, *87*(5), 1617–1619. doi:10.1016/j.mee.2009.10.037

Hassanin, H., Modica, F., El-Sayed, M. A., Liu, J., & Essa, K. (2016). Manufacturing of Ti–6Al–4V Micro-Implantable Parts Using Hybrid Selective Laser Melting and Micro-Electrical Discharge Machining. *Advanced Engineering Materials, 18*(9), 1544–1549. doi:10.1002/adem.201600172

Hassanin, H., Zweiri, Y., Finet, L., Essa, K., Qiu, C., & Attallah, M. (2021). Laser Powder Bed Fusion of Ti-6Al-2Sn-4Zr-6Mo Alloy and Properties Prediction Using Deep Learning Approaches. *Materials, 14*(8), 2056. https://www.mdpi.com/1996-1944/14/8/2056

Hinton, G. E., Osindero, S., & Teh, Y.-W. (2006). A Fast Learning Algorithm for Deep Belief Nets. *Neural Computation, 18*(7), 1527–1554. doi:10.1162/neco.2006.18.7.1527%M16764513

Hinton, G. E., & Salakhutdinov, R. R. (2006). Reducing the Dimensionality of Data with Neural Networks. *Science, 313*(5786), 504–507. doi:10.1126cience.1127647 PMID:16873662

Jiménez, A., Bidare, P., Hassanin, H., Tarlochan, F., Dimov, S., & Essa, K. (2021). Powder-based laser hybrid additive manufacturing of metals: A review. *International Journal of Advanced Manufacturing Technology, 114*(1-2), 63–96. Advance online publication. doi:10.100700170-021-06855-4

Kayacan, M. Y., Özsoy, K., Duman, B., Yilmaz, N., & Kayacan, M. C. (2019). A study on elimination of failures resulting from layering and internal stresses in Powder Bed Fusion (PBF) additive manufacturing. *Materials and Manufacturing Processes, 34*(13), 1467–1475. doi:10.1080/10426914.2019.1655151

Klippstein, H., Hassanin, H., Diaz De Cerio Sanchez, A., Zweiri, Y., & Seneviratne, L. (2018). Additive Manufacturing of Porous Structures for Unmanned Aerial Vehicles Applications. *Advanced Engineering Materials, 20*(9), 1800290. doi:10.1002/adem.201800290

Langford, T., Mohammed, A., Essa, K., Elshaer, A., & Hassanin, H. (2021). 4D Printing of Origami Structures for Minimally Invasive Surgeries Using Functional Scaffold. *Applied Sciences, 11*(1), 332. https://www.mdpi.com/2076-3417/11/1/332

Mohammed, A., Elshaer, A., Sareh, P., Elsayed, M., & Hassanin, H. (2020). Additive Manufacturing Technologies for Drug Delivery Applications. *International Journal of Pharmaceutics, 580*, 119245. doi:10.1016/j.ijpharm.2020.119245 PMID:32201252

Okafor-Muo, O. L., Hassanin, H., Kayyali, R., & ElShaer, A. (2020). 3D Printing of Solid Oral Dosage Forms: Numerous Challenges With Unique Opportunities. *Journal of Pharmaceutical Sciences*, *109*(12), 3535–3550. doi:10.1016/j.xphs.2020.08.029 PMID:32976900

Olson, G. B. (1997). Computational Design of Hierarchically Structured Materials. *Science*, *277*(5330), 1237–1242. doi:10.1126cience.277.5330.1237

Sabouri, A., Yetisen, A. K., Sadigzade, R., Hassanin, H., Essa, K., & Butt, H. (2017). Three-Dimensional Microstructured Lattices for Oil Sensing. *Energy & Fuels*, *31*(3), 2524–2529. doi:10.1021/acs.energyfuels.6b02850

Scherillo, F., Astarita, A., Carrino, L., Pirozzi, C., Prisco, U., & Squillace, A. (2019). Linear friction welding of Ti-6Al-4V parts produced by electron beam melting. *Materials and Manufacturing Processes*, *34*(2), 201–207. doi:10.1080/10426914.2018.1532086

Schmitt, M., Mehta, R. M., & Kim, I. Y. (2020). Additive manufacturing infill optimization for automotive 3D-printed ABS components. *Rapid Prototyping Journal*, *26*(1), 89–99. doi:10.1108/RPJ-01-2019-0007

Tamburrino, F., Graziosi, S., & Bordegoni, M. (2018). The design process of additively manufactured mesoscale lattice structures: A review. *Journal of Computing and Information Science in Engineering*, *18*(4), 040801. Advance online publication. doi:10.1115/1.4040131

Tan, X. P., Tan, Y. J., Chow, C. S. L., Tor, S. B., & Yeong, W. Y. (2017). Metallic powder-bed based 3D printing of cellular scaffolds for orthopaedic implants: A state-of-the-art review on manufacturing, topological design, mechanical properties and biocompatibility. *Materials Science and Engineering C*, *76*, 1328–1343. doi:10.1016/j.msec.2017.02.094 PMID:28482501

Weichert, D., Link, P., Stoll, A., Rüping, S., Ihlenfeldt, S., & Wrobel, S. (2019). A review of machine learning for the optimization of production processes. *International Journal of Advanced Manufacturing Technology*, *104*(5-8), 1889–1902. Advance online publication. doi:10.100700170-019-03988-5

Yan, C., Hao, L., Hussein, A., & Raymont, D. (2012). Evaluations of cellular lattice structures manufactured using selective laser melting. *International Journal of Machine Tools & Manufacture*, *62*, 32–38. doi:10.1016/j.ijmachtools.2012.06.002

Yang, E., Leary, M., Lozanovski, B., Downing, D., Mazur, M., Sarker, A., Khorasani, A. M., Jones, A., Maconachie, T., Bateman, S., Easton, M., Qian, M., Choong, P., & Brandt, M. (2019). Effect of geometry on the mechanical properties of Ti-6Al-4V Gyroid structures fabricated via SLM: A numerical study. *Materials & Design*, *184*, 108165. doi:10.1016/j.matdes.2019.108165

Yang, J., Li, S., Wang, Z., Dong, H., Wang, J., & Tang, S. (2020). Using deep learning to detect defects in manufacturing: A comprehensive survey and current challenges. *Materials (Basel)*, *13*(24), 1–23. doi:10.3390/ma13245755 PMID:33339413

Yang, L., Harrysson, O., Cormier, D., West, H., Gong, H., & Stucker, B. (2015). Additive Manufacturing of Metal Cellular Structures: Design and Fabrication. *JOM*, *67*(3), 608–615. doi:10.100711837-015-1322-y

Yuan, B., Guss, G. M., Wilson, A. C., Hau-Riege, S. P., DePond, P. J., McMains, S., Matthews, M. J., & Giera, B. (2018). Machine-Learning-Based Monitoring of Laser Powder Bed Fusion. *Advanced Materials Technologies*, *3*(12), 1800136. doi:10.1002/admt.201800136

Zhang, X. Z., Leary, M., Tang, H. P., Song, T., & Qian, M. (2018). Selective electron beam manufactured Ti-6Al-4V lattice structures for orthopedic implant applications: Current status and outstanding challenges. *Current Opinion in Solid State and Materials Science*, *22*(3), 75–99. doi:10.1016/j.cossms.2018.05.002

# Chapter 3
# Regression and Artificial Intelligence Models to Predict the Surface Roughness in Additive Manufacturing

**Mohamed Hamoud Ahmed**
*Helwan University, Egypt*

**Azza fathallah Barakat**
*Helwan University, Egypt*

**Abuubakr Ibrahim Abdelwahab**
*Helwan University, Egypt*

## ABSTRACT

*In additive manufacturing (AM), it is necessary to study the surface roughness, which affected the building parameters such as layer thickness and building orientation. Some AM machines have minimum layer thickness that doesn't satisfy the desired roughness. Also, it produces a fine surface that isn't required. This increases the building time and cost without any benefits. To overcome these problems and achieve a certain surface roughness, a prediction model is proposed in this chapter. Regression models were used to predict the surface roughness through the building orientation. ANN was used to predict the surface roughness through the building orientation and the layer thickness together. ANN was constructed based on experimental work that study the effect of layer thickness and building orientation on the surface roughness. Some data were used in the training process and others were used in the verification process. The results show that the layer thickness parameter has an effect more than the building orientation parameter on the surface roughness.*

DOI: 10.4018/978-1-7998-8516-0.ch003

## INTRODUCTION

Fused Deposition Modeling (FDM) is an Additive Manufacturing (AM) technique that is commonly call as three-dimensional printing (3DP) technology. In this technology, a filament made of different types of materials such as PLA, acrylonitrile butadiene styrene (ABS), etc. is melted and extruded through an extruder onto a print bed. parts are built by depositing the filament, layer by layer.(Kandananond, 2020). Because this technology is being increasingly used many applications, the technology is also known by several other names like Rapid Manufacturing (RM), Rapid Prototyping (RP), Layered Manufacturing (LM), digital fabrication, Free Form Fabrication (FFF), Desktop Manufacturing, Direct CAD Manufacturing, and Computer Automated Manufacturing (Hamoud, 2018a). One recent aspect of product data communication is the communication standard between CAD systems and solid free form fabrication (SFF) systems. Here, a format called STL (Standard Transform Language) file. In commercial AM systems, the process begins by slicing the CAD model to obtain a 2D contour at each level of the build axis (Z-axis). Starting from the base 2D contour, slice thickness is defined by the user which added cumulatively at successive slicing planes (Sobhi et al., 2019).

## BACKGROUND

Surface Roughness is important area studied in AM. Recently, the investigations pay their attention in improving the produced surface roughness which depends on the process parameters such as building orientation and layer thickness during the part building (Hamoud, 2018b). Karin Kandananond (Kandananond, 2020) study the effect of three Process parameters which were nozzle temperature, bed temperature, and printing speed, on the surface roughness of the specimens. Roughness measurements were made on the flat surfaces at the top and bottom of the specimens. Response surface method (RSM), was utilized in this study to get the optimum parameters. The results indicated that the bed temperature and the printing speed had a significant impact on the surface roughness. In addition, there was a non-linear relationship between the bed temperature and the surface roughness. The optimal process parameters in this study can use as guidelines for the practitioners to achieve the highest performance when they use ABS material. Esther Molero et. al. (Molero et al., 2020) studied the effect of five parameters namely layer height, extrusion temperature, print speed, print acceleration and flow on the surface roughness for PLA material on FDM machine. ten data mining algorithms were used to get predication models. Mohd Nazri Ahmad, et. al. (Ahmad et al., 2019) used a Taguchi Method for the optimization of the product or process quality. In this work, an orthogonal array of L9 ($3^4$) was used to determine four parameters with three levels each. The samples with ASTM D638 type IV standard ware fabricated by 3D Printer type FDM. Thus, the result shows the optimum parameters are print pattern (cross), orientation on Y-axis (0°), support angle (0°) and side walk (0.15 mm). Mercedes Pérez, et. al. (Pérez et al., 2018) this research aims at recognition of critical parameters that cause a reduction surface roughness in FDM technique. Five process parameters were studied. The results showed that there are two parameters that controlling the surface roughness namely layer height and wall thickness, while the remaining parameters namely printing path, printing speed, and temperature showed no clear influence on surface roughness. Ebrahim Vahabli, et. al. (Vahabli & Rahmati, 2017) this study developed a model based on empirical data based on optimized artificial neural networks (ANNs) to estimate the surface roughness distribution in fused deposition modelling (FDM) parts. The optimized parameters were time, cost and quality. several improvements were presented in surface roughness distribution modelling including a more suitable method for process parameter selection according to the design criteria and improvements in the overall surface roughness of parts as compared toanalytical methods. GAJDOŠ Ivan et. al. (Gajdos et al., 2015) the aim of this work is to improve the surface finish in FDM technique. specimens were blasted with sodium bicarbonate and glass beads. The values of Ra and Rz were measured on the spacemen's and compared the impact of selected pre-treatment, blasting medium and blasting conditions on surface finish of FDM prototypes. The results show that the surface pre-treatment with NanoSeal 180W or spraying putty have even better results compared to blasted surface. Alberto Boschetto et. al. (Boschetto et al., 2013) present a feed-forward neural network to fit experimental data in order to determine roughness parameters models entire part surface. Daekeon Ahn and et. al. (Ahn et al., 2011) the authors investigate the relation between surface roughness and overlap interval is based on a surface roughness formulation in FDM. Additionally, effects of surface angle and filament shape are analyzed and discussed to predict surface roughness distribution by the overlap interval variation. The previous work proves the important of studying the parameters that effect the surface roughness. this chapter aims at studying the effect of process parameters namely layer thickness and building orientation on the surface roughness and investigate predication models based on the experimental study. Regression models were investigated firstly to predict the surface roughness through the building orientation. In addition, ANN was used to predict the surface roughness through the building orientation and the layer thickness together. ANN was constructed based on experimental work that study the effect of layer thickness and building orientation on the surface roughness. Some data were used in the training process and other were used in the verification process. The experimental procedure and the results and the prediction models will be discussing in the following sections.

# DESIGN OF EXPERIMENT

This section describes the technical specifications of the AM machine used in the experiment and the influence of its parameters on the surface roughness. The AM machine is MakerBot Replicator 2X Experimental 3D Printer as shown in Figure 1 with technical specifications presented in Table 1. There are two process parameters to be considered in the experiment. The parameters are the layer thickness and the building orientation. The available layer thickness on the machine ranges from 0.1 to 0.4 mm in step of 0.01mm. Therefore, there are 30 values for the layer thickness. The available angles for reorient the 3D CAD model ranging from 0 (designed orientation) to 90° in step of 1°. Therefore, there are 90 values for the building orientation. To get the greatest coverage of decision space and the tangible effect on surface roughness, the layer thickness values were taken at minimum value (0.1 mm) and maximum value (0.4 mm) and in between like (0.3 mm). On the other hand, the building angles were ranged from 0° to 80° in step of 10°.

## Design of 3D CAD Specimen

In this experiment, it is needed to build nine specimens for each value of layer thickness. That means it is required to build 27 specimens. That is why a new special CAD specimen was designed to reduce the number of rebuilding the specimens. The designed specimen is shown in Figure 2. The physical specimens are represented in Figure 3.

## Surface Roughness Assessments

The device used in the experiment was surtronic 3+ by Taylor Hobson precision company as shown in Figure 4. The cut off length (Lc) was 0.8 mm and the evaluation length (Ln) was 5*Lc, which mean the evaluation length was 4 mm. The surface parameter chosen to assessment the surface was Ra – Arithmetic mean surface roughness for the following reasons:

- The most commonly used parameter to monitor a production process.
- Default parameter on a drawing if not otherwise specified.
- Available even in the least sophisticated instruments.

The device was calibrated firstly with the test piece to grantee the measured data.

*Figure 1. Makerbot replicator 2x experimental 3D printer*

*Table 1. Technical specifications of the 3D printer*

| | |
|---|---|
| Print technology | Fused Deposition Modeling |
| Build volume | 24.6 l x 15.2 w x 15.5 h cm |
| Minimum layer height | 100 microns |
| Filament diameter | 1.75 mm |
| Filament material | ABS |
| Nozzle diameter | 0.4 mm |
| File types | Stl, obj, thing |
| Connectivity | USB, SD card (both included) |

*Figure 2. CAD model of the designed specimen*

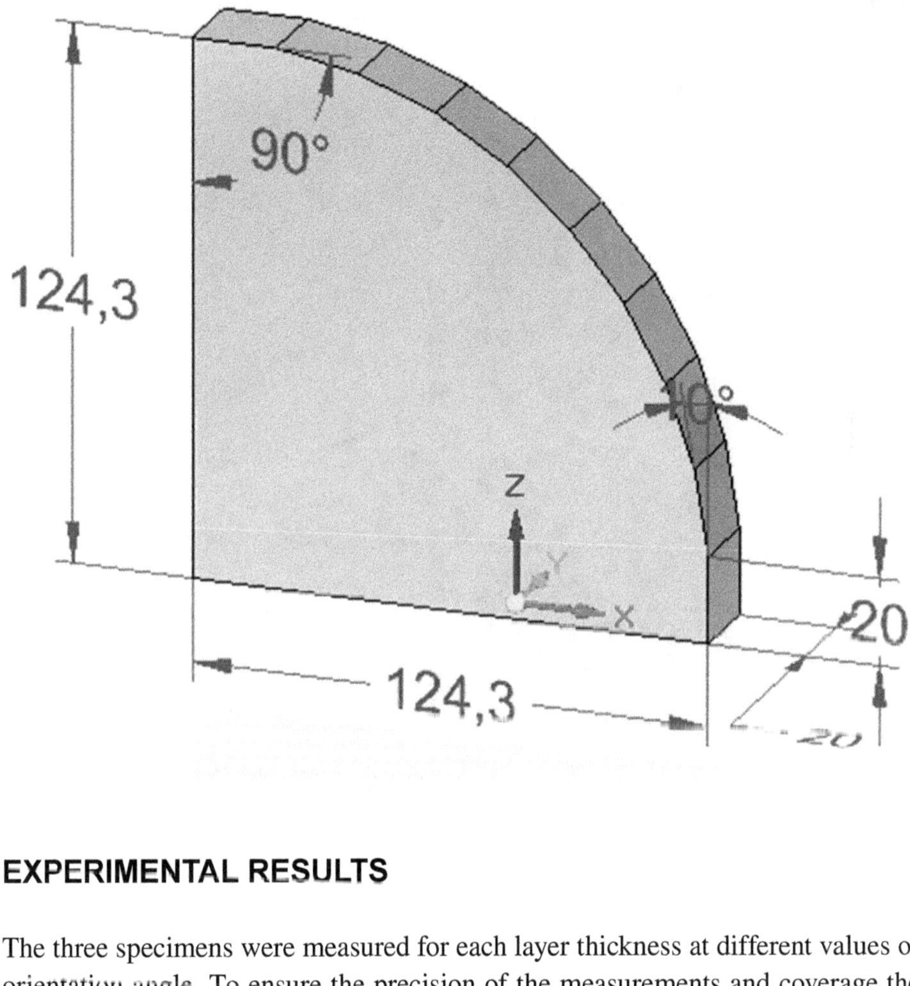


## EXPERIMENTAL RESULTS

The three specimens were measured for each layer thickness at different values of orientation angle. To ensure the precision of the measurements and coverage the surface area, there were three values for each reading data in different position on the measuring surface. After that, the average value was calculated to evaluate the surface roughness. Table 2, Table 3, and Table 4 show the results of the measurements for three levels of layer thickness. The relationship between the building orientation and the surface roughness at three different values of layer thickness is shown in Figure 5.

*Figure 3. Specimen building using 0.4 mm layer thickness*

*Figure 4. Specimen building using 0.3 mm layer thickness*

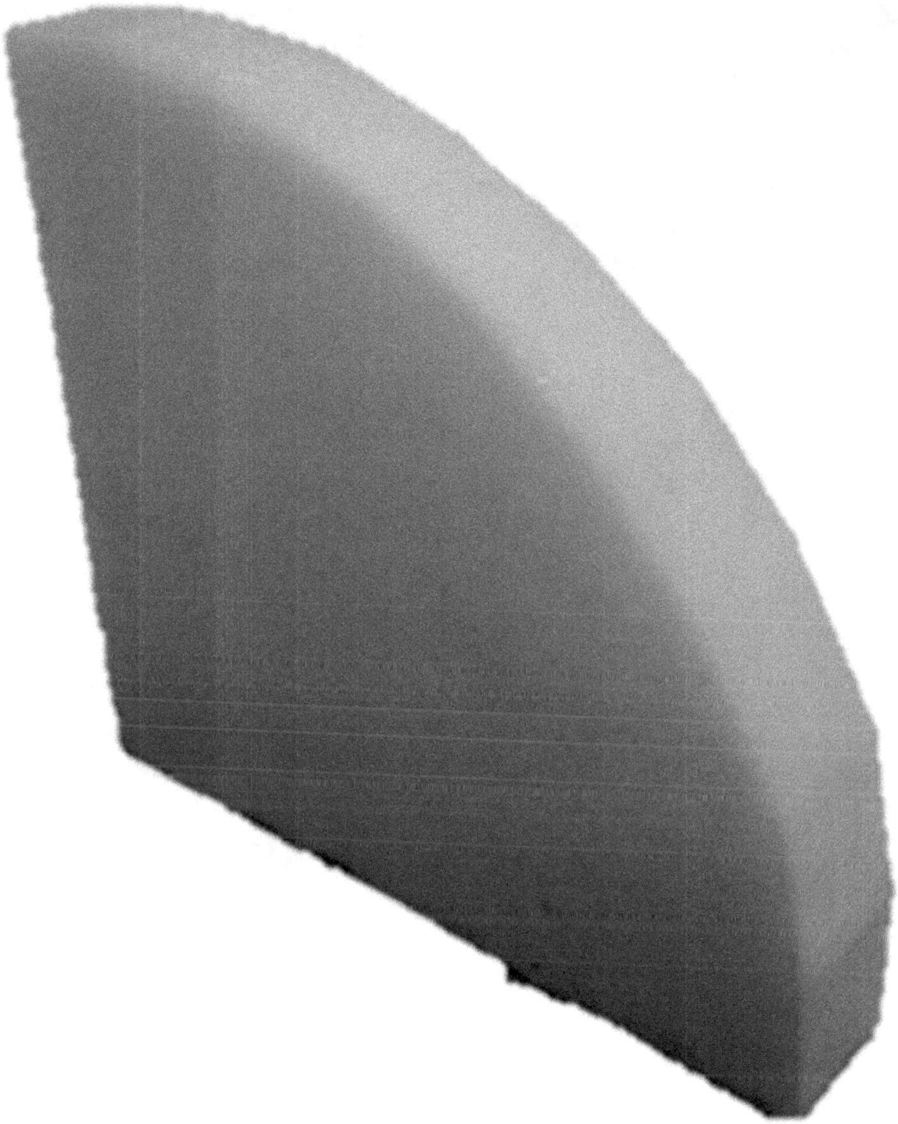

## POLYNOMIAL REGRESSION MODEL

Polynomials are very flexible and useful where a model must be developed empirically. In the presented model, polynomial regression fits a nonlinear relationship between

*Figure 5. Specimen building using 0.1 mm layer thickness*

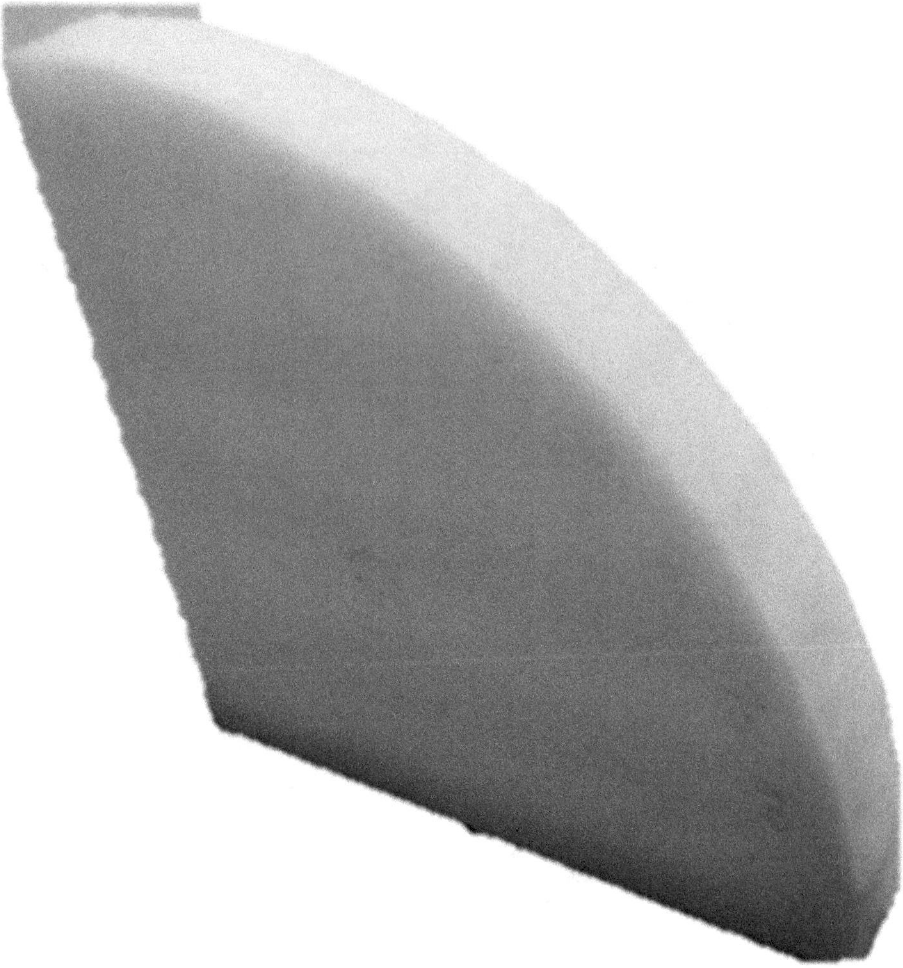

the value of building orientation and the corresponding surface roughness at certain layer thickness. The general form of the polynomial is

$$f(x) = a_{n-1} x^{n-1} + a_{n-2} x^{n-2} + \ldots\ldots\ldots\ldots + a_2 x^2 + a_1 x + a_0;$$

where n is the number of measuring points, f(x) is the surface roughness (Ra), x is the rotation angle (O) and a is the coefficients of polynomial. In this study n=6 to get the prediction model, the average values of roughness token at O = 10°, 20°, 40°, 50°, 70°, 80° and the remaining values at O = 30°, 60° used for the model verifications.

*Figure 6. Experiment setup using surtronic 3+*

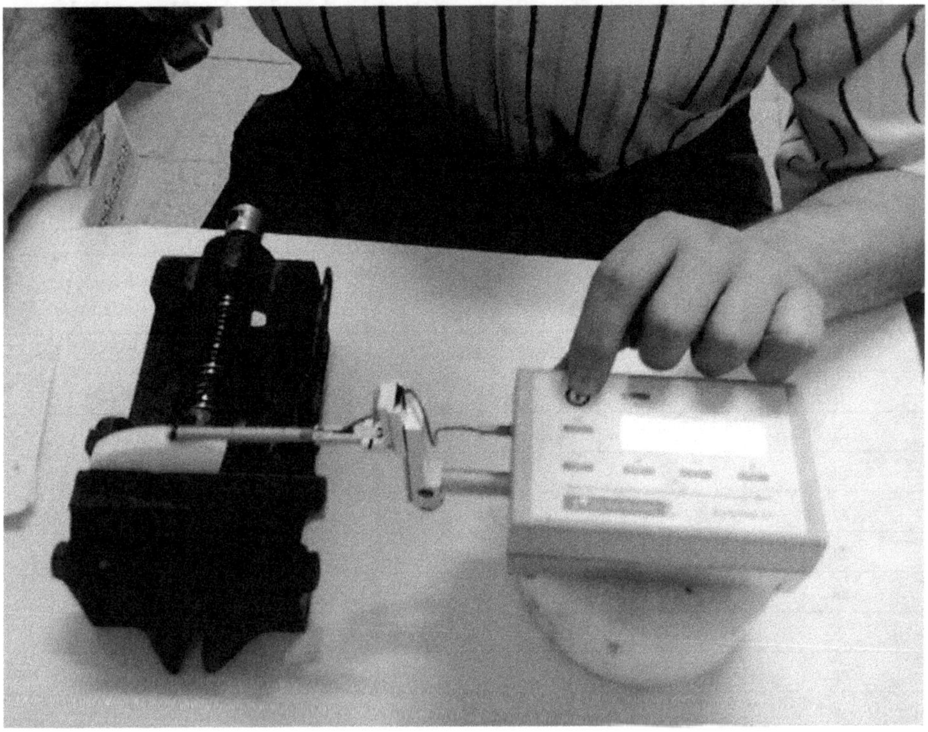

For layer thickness = 0.1mm

$$Ra = 2.36 \times 10^{-7} O^4 - 2.66*10^{-05} O^3 + 0.0004 O^2 + 0.094 O + 6.103 \quad (1)$$

For layer thickness = 0.3mm

$$Ra = -3.013 \times 10^{-6} O^4 +0.00057 O^3 -0.0362 O^2 +1.0328 O +12.93 \quad (2)$$

For layer thickness = 0.4mm

$$Ra = 5.645 \times 10^{-7} O^5 -0.00012 O^4 +0.0099 O^3 -0.338 O^2 +5.264 O - 4.138 \quad (3)$$

## a. Model Verification

Figure 6, Figure 7, and Figure 8 show the relationship between the experimental measurements data and the predictive model. Table 5, Table 6, and Table 7 represent the different between the predictive model and the actual measurements. For verification, the model tested at O = 30° and 60° at the three levels of layer thickness. The results represented in Table 8.

*Table 2. Raw data and average values of surface roughness at layer thickness = 0.1 mm*

| Surface angle (O) in degree | Surface Roughness (Ra) in Micron | | | Average Roughness |
|---|---|---|---|---|
| | 1st run | 2nd run | 3rd run | |
| 0 | 6.4 | 6.4 | 6.6 | 6.47 |
| 10 | 6.2 | 7.8 | 7.2 | 7.07 |
| 20 | 8.4 | 8.2 | 7.4 | 8 |
| 30 | 8.9 | 9 | 8.5 | 8.8 |
| 40 | 9.4 | 9.7 | 9.5 | 9.53 |
| 50 | 9.4 | 10.6 | 10.4 | 10.13 |
| 60 | 11 | 10.8 | 10.6 | 10.8 |
| 70 | 11.6 | 12 | 11 | 11.53 |
| 80 | 12.2 | 13 | 12.8 | 12.67 |

*Table 3. Raw data and average values of surface roughness at layer thickness = 0.3 mm*

| Surface angle (O) in degree | Surface Roughness (Ra) in Micron | | | Average Roughness |
|---|---|---|---|---|
| | 1st run | 2nd run | 3rd run | |
| 0 | 20.2 | 19.8 | 16 | 18.67 |
| 10 | 19.8 | 20.6 | 20.2 | 20.2 |
| 20 | 23 | 23.4 | 23.2 | 23.2 |
| 30 | 24.2 | 23.4 | 24.5 | 24.03 |
| 40 | 25.2 | 25.6 | 26.4 | 25.73 |
| 50 | 27.8 | 26.9 | 27.6 | 27.43 |
| 60 | 28.6 | 28.8 | 29.8 | 29.07 |
| 70 | 34.2 | 33.2 | 34.6 | 34 |
| 80 | 35.8 | 36.9 | 37.4 | 36.7 |

*Table 4. Raw data and average values of surface roughness at layer thickness = 0.4 mm*

| Surface angle (O) in degree | Surface Roughness (Ra) in Micron | | | Average Roughness |
|---|---|---|---|---|
| | 1st run | 2nd run | 3rd run | |
| 0 | 22.8 | 21 | 23 | 22.27 |
| 10 | 22.8 | 23.8 | 23.6 | 23.4 |
| 20 | 26.4 | 27.4 | 27.2 | 27 |
| 30 | 31.5 | 33 | 33.9 | 32.8 |
| 40 | 38 | 38.6 | 37.8 | 38.13 |
| 50 | 48.2 | 47.6 | 49 | 46.7 |
| 60 | 51.8 | 55.8 | 55 | 54.2 |
| 70 | 58.9 | 60.4 | 60.9 | 60.07 |
| 80 | 65.4 | 69 | 65 | 66.47 |
| 90 | 63 | 70 | 61 | 64.67 |

*Figure 7. Effect of the building orientation and layer thickness on the surface roughness*

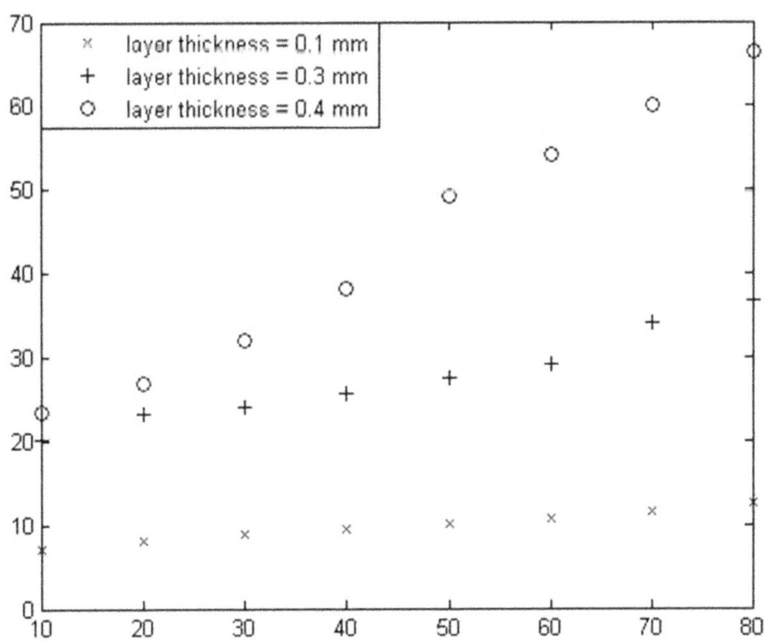

*Figure 8. Predictive model at layer thickness = 0.1 mm*

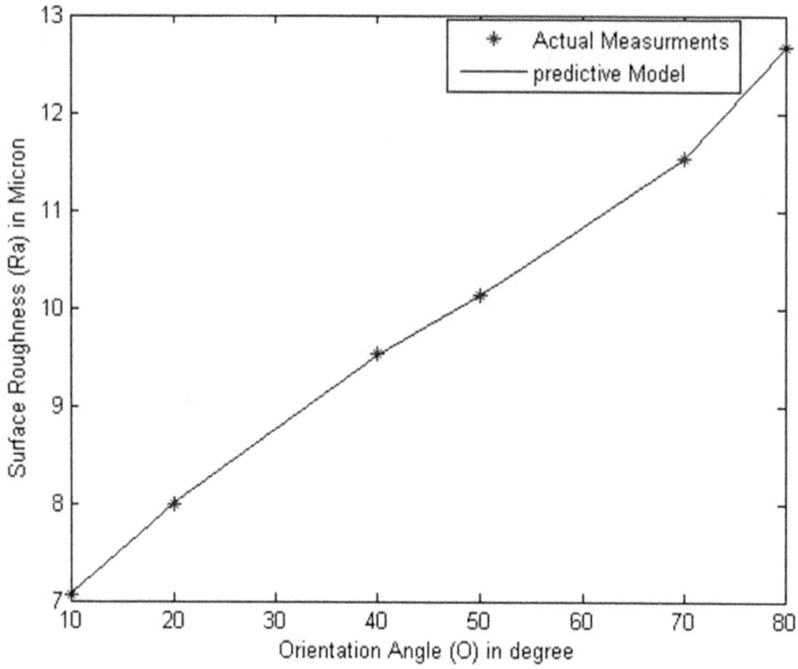


*Table 5. The differences between the measured values and the predictive values at layer thickness = 0.1 mm*

| Orientation angle (degree) | Actual roughness (micron) | Predictive roughness (micron) | Differences (micron) |
|---|---|---|---|
| 10 | 7.0700 | 7.0693 | 0.0007 |
| 20 | 8.0000 | 8.0020 | -0.0020 |
| 40 | 9.5300 | 9.5249 | 0.0051 |
| 50 | 10.1300 | 10.1351 | -0.0051 |
| 70 | 11.5300 | 11.5280 | 0.0020 |
| 80 | 12.6700 | 12.6707 | -0.0007 |

*Figure 9. Predictive model at layer thickness = 0.3 mm*

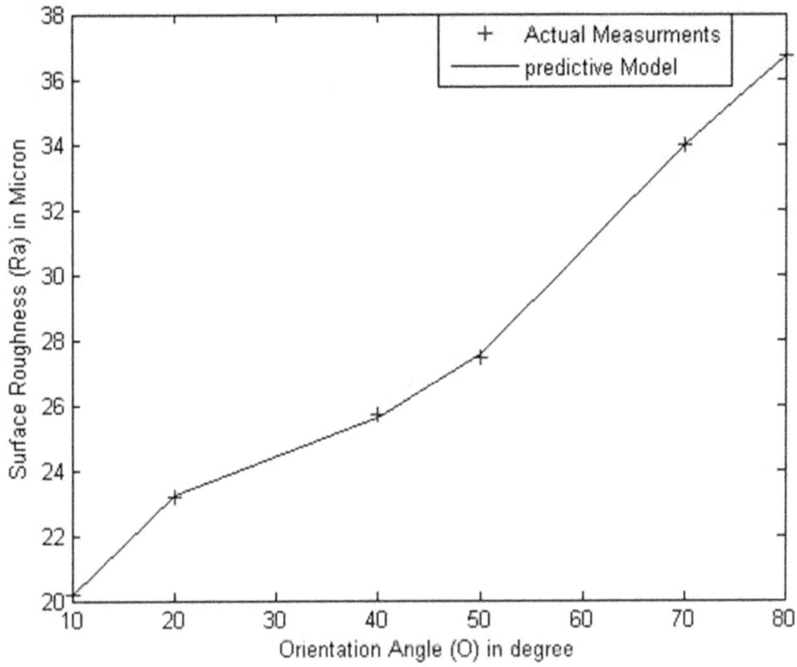


*Table 6. The differences between the measured values and the predictive values at layer thickness = 0.3 mm*

| Orientation angle (degree) | Actual roughness (micron) | Predictive roughness (micron) | Differences (micron) |
|---|---|---|---|
| 10 | 20.2000 | 20.1841 | 0.0159 |
| 20 | 23.2000 | 23.2445 | -0.0445 |
| 40 | 25.7300 | 25.6187 | 0.1113 |
| 50 | 27.4300 | 27.5413 | -0.1113 |
| 70 | 34.0000 | 33.9555 | 0.0445 |
| 80 | 36.7000 | 36.7159 | -0.0159 |

*Figure 10. Predictive model at layer thickness = 0.4 mm*

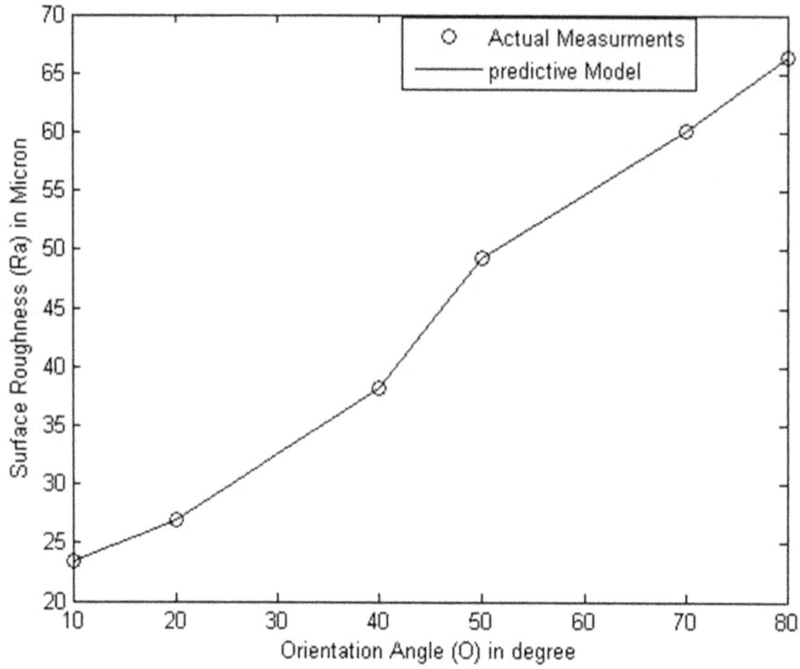


*Table 7. The differences between the measured values and the predictive values at layer thickness = 0.4 mm*

| Orientation angle (degree) | Actual roughness (micron) | Predictive roughness (micron) | Differences (micron) $*10^{-12}$ |
|---|---|---|---|
| 10 | 23.4000 | 23.4000 | -0.4974 |
| 20 | 27.0000 | 27.0000 | -0.1137 |
| 40 | 38.1300 | 38.1300 | -0.0782 |
| 50 | 46.700 | 46.700 | 0.1350 |
| 70 | 60.0700 | 60.0700 | -0.1492 |
| 80 | 66.4700 | 66.4700 | -0.2558 |

*Table 8. The results of model validation*

| Angle (deg) | Layer thickness (mm) | | | | | | | | |
|---|---|---|---|---|---|---|---|---|---|
| | 0.1 | | | 0.3 | | | 0.4 | | |
| | Measured | Predictive | Error | Measured | Predictive | Error | Measured | Predictive | Error |
| 30 | 8.8 | 8.83 | -0.03 | 24.03 | 24.2 | -0.17 | 32.8 | 29.7 | 3.1 |
| 60 | 10.8 | 10.75 | 0.05 | 29.07 | 30.2 | -1.13 | 54.2 | 54.61 | -0.41 |

## ARTIFICIAL NEURAL NETWORK (ANN) Model

Some AM machines have many different values of layer thickness and building orientation. That is why ANN approach is more suitable than a statistical approach to predict the surface roughness. An ANN can learn to find the solution and can predict the missing or incomplete data. The main idea of the ANN is to create an algorithm that simulates the human brain thinking in some features. Neural networks are typically organized in layers. Layers are made up of a number of interconnected 'nodes' which contain an 'activation function'. Patterns are presented to the network via the 'input layer', which communicates to one or more 'hidden layers' where the actual processing is done via a system of weighted 'connections. The hidden layers then link to an 'output layer' wherc the answer is output as shown in Figure 9.

*Figure 11. Neural network architecture*

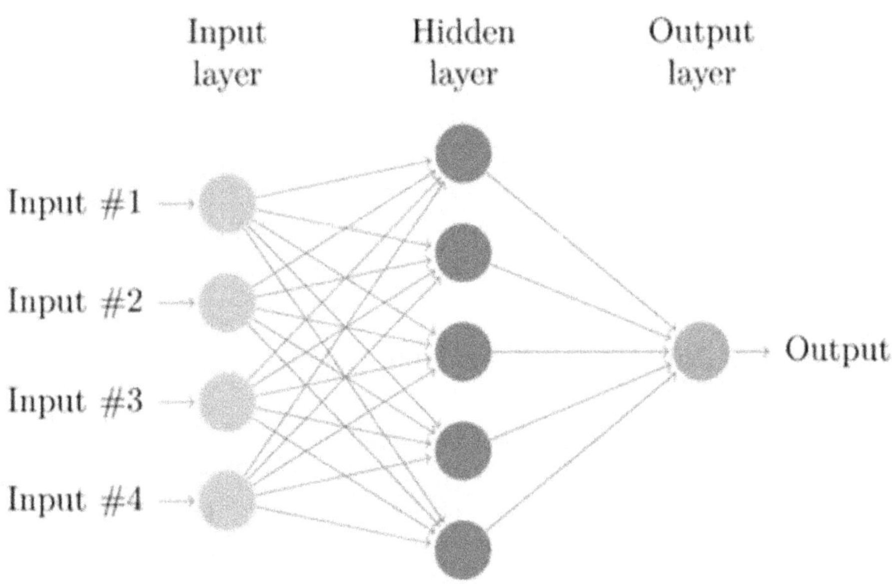

*Table 9. Total number of experimental data*

| Experiment # | Layer Thickness | Orientation | Ra |
|---|---|---|---|
| 1 | 0.1 | 0 | 6.47 |
| 2 | 0.1 | 10 | 7.07 |
| 3 | 0.1 | 20 | 8 |
| 4 | 0.1 | 30 | 8.8 |
| 5 | 0.1 | 40 | 9.53 |
| 6 | 0.1 | 50 | 10.1 |
| 7 | 0.1 | 60 | 10.8 |
| 8 | 0.1 | 70 | 11.5 |
| 9 | 0.1 | 80 | 12.7 |
| 10 | 0.3 | 0 | 18.7 |
| 11 | 0.3 | 10 | 20.2 |
| 12 | 0.3 | 20 | 23.2 |
| 13 | 0.3 | 30 | 24 |
| 14 | 0.3 | 40 | 25.7 |
| 15 | 0.3 | 50 | 27.4 |
| 16 | 0.3 | 60 | 29.1 |
| 17 | 0.3 | 70 | 34 |
| 18 | 0.3 | 80 | 36.7 |
| 19 | 0.4 | 0 | 22.3 |
| 20 | 0.4 | 10 | 23.4 |
| 21 | 0.4 | 20 | 27 |
| 22 | 0.4 | 30 | 32.8 |
| 23 | 0.4 | 40 | 38.1 |
| 24 | 0.4 | 50 | 46.7 |
| 25 | 0.4 | 60 | 54.2 |
| 26 | 0.4 | 70 | 60.1 |
| 27 | 0.4 | 80 | 66.5 |

## Experimental Data for Prediction using ANN

ANN is an approach capable of identifying the influence of layer thickness and building orientation as inputs on surface quality as output. The trained neural net can be used as a predicative model for surface roughness. The first step in applying the ANN is dividing the total experimental data (Table 9) into validation data (Table 10) and training data (Table 11).

*Table 10. Validation data*

| Experiment # | Layer Thickness | Orientation | Ra |
|---|---|---|---|
| 4 | 0.1 | 30 | 8.8 |
| 8 | 0.1 | 70 | 11.5 |
| 14 | 0.3 | 40 | 25.7 |
| 17 | 0.3 | 70 | 34 |
| 22 | 0.4 | 30 | 32.8 |
| 24 | 0.4 | 50 | 46.7 |

*Table 11. Training data*

| Experiment# | Layer Thickness | Orientation | Ra |
|---|---|---|---|
| 1 | 0.1 | 0 | 6.47 |
| 2 | 0.1 | 10 | 7.07 |
| 3 | 0.1 | 20 | 8 |
| 5 | 0.1 | 40 | 9.53 |
| 6 | 0.1 | 50 | 10.1 |
| 7 | 0.1 | 60 | 10.8 |
| 9 | 0.1 | 80 | 12.7 |
| 10 | 0.3 | 0 | 18.7 |
| 11 | 0.3 | 10 | 20.2 |
| 12 | 0.3 | 20 | 23.2 |
| 13 | 0.3 | 30 | 24 |
| 15 | 0.3 | 50 | 27.4 |
| 16 | 0.3 | 60 | 29.1 |
| 18 | 0.3 | 80 | 36.7 |
| 19 | 0.4 | 0 | 22.3 |
| 20 | 0.4 | 10 | 23.4 |
| 21 | 0.4 | 20 | 27 |
| 23 | 0.4 | 40 | 38.1 |
| 25 | 0.4 | 60 | 54.2 |
| 26 | 0.4 | 70 | 60.1 |
| 27 | 0.4 | 80 | 66.5 |

*Table 12. Results summary for different design of ANN*

| No. of hidden layers | Network structure design | Actual training cycles | Error | | | | | | |
|---|---|---|---|---|---|---|---|---|---|
| | | | Training | | | Test | | | |
| | | | Max. | Avg. | Min. | Max. | Avg. | Min. | MSE |
| 1 | 2*1*1 | $6*10^6$ | 0.015 | $4*10^{-3}$ | $6*10^{-6}$ | 6.386 | 3.27 | 1.336 | 13.834 |
| 1 | 2*2*1 | $6*10^6$ | $1*10^{-3}$ | $3*10^{-4}$ | $4*10^{-6}$ | 3.635 | 2.368 | 0.876 | 6.4687 |
| 1 | 2*3*1 | $6*10^6$ | $4*10^{-4}$ | $6*10^{-5}$ | 0 | 1.789 | 0.93 | 0.131 | 1.3563 |
| 1 | 2*4*1 | $6*10^6$ | $2*10^{-5}$ | $6*10^{-6}$ | 0 | 13.209 | 3.814 | 0.077 | 37.288 |
| 2 | 2*1*1*1 | $6*10^6$ | 0.014 | 0.004 | 0 | 6.129 | 2.631 | 0.189 | 11.137 |
| 2 | 2*3*3*1 | 58801 | 0.0027 | $5*10^{-4}$ | $2*10^{-6}$ | 2.308 | 0.919 | 0.177 | 1.4220 |
| 2 | 2*4*3*1 | 5096301 | $4*10^{-5}$ | $1*10^{-5}$ | $2*10^{-8}$ | 2.723 | 0.736 | 0.175 | 1.3497 |
| 2 | 2*3*5*1 | $6*10^6$ | $1*10^{-6}$ | $1*10^{-7}$ | 0 | 10.759 | 2.948 | 0.03 | 22.058 |
| 2 | 2*5*4*1 | 28101 | $5*10^{-4}$ | $7*10^{-5}$ | 0 | 1.807 | 0.841 | 0.096 | 1.2309 |
| 3 | 2*2*2*2*1 | $6*10^6$ | $1*10^{-3}$ | $4*10^{-4}$ | $1*10^{-8}$ | 3.882 | 2.335 | 0.482 | 6.8683 |
| 3 | 2*3*3*3*1 | $6*10^6$ | $1*10^{-6}$ | $2*10^{-7}$ | 0 | 3.442 | 1.431 | 0.075 | 3.765 |
| 3 | 2*4*3*2*1 | 3301 | $1*10^{-3}$ | $1*10^{-4}$ | $3*10^{-7}$ | 1.750 | 1.015 | 0.434 | 1.297 |
| 3 | 2*4*3*3*1 | $6*10^6$ | $3*10^{-4}$ | $5*10^{-4}$ | $1*10^{-8}$ | 1.982 | 0.997 | 0.247 | 1.394 |

## Training Process for ANN

In the training process, the samples are presented to the network one by one. Using the current weights, the network predicts a corresponding surface roughness for a given input set. The difference between the actual and predicted output for each input and output set is calculated and fed back to adjust the weights of network using the standard back propagation algorithm. The new weights are then used to calculate the output of the next input set. The training process continues until all actual and predicted outputs are close enough to one another. In this implementation, the target error used for training the network was 0.01μmm at $6*10^6$ cycles. Table 12 shows the number of network trails to determine the best structure that satisfy the minimum Mean Square Error (MSE). From these trails, it is clear there are three minimum values of MSE at 2*4*3*1, 2*5*4*1, and 2*4*3*2*1 respectively. Figure 10 shows the training diagrams of the three structures.

*Figure 12. Learning progress for 2*4*3*2*1 ANN structure*

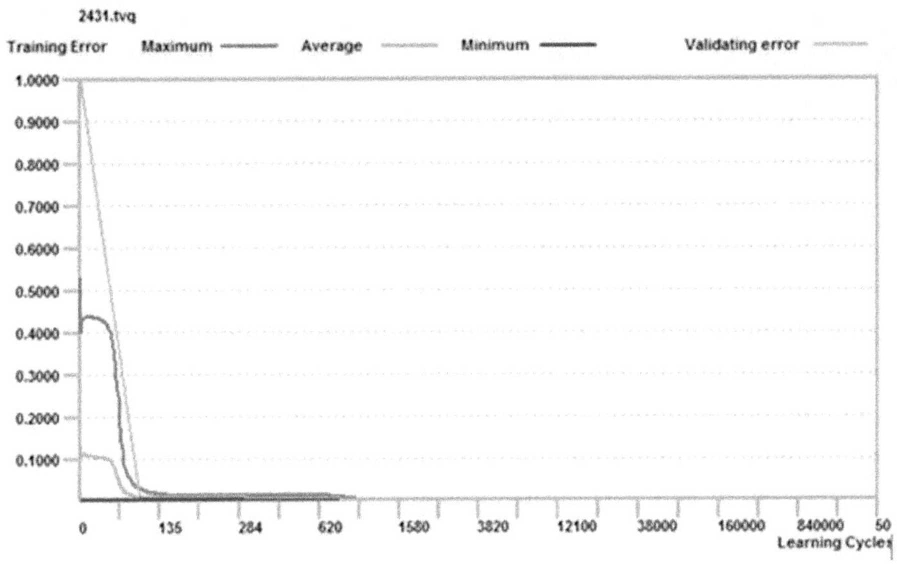


*Figure 13. Learning progress for 2*5*4*1 ANN structure*

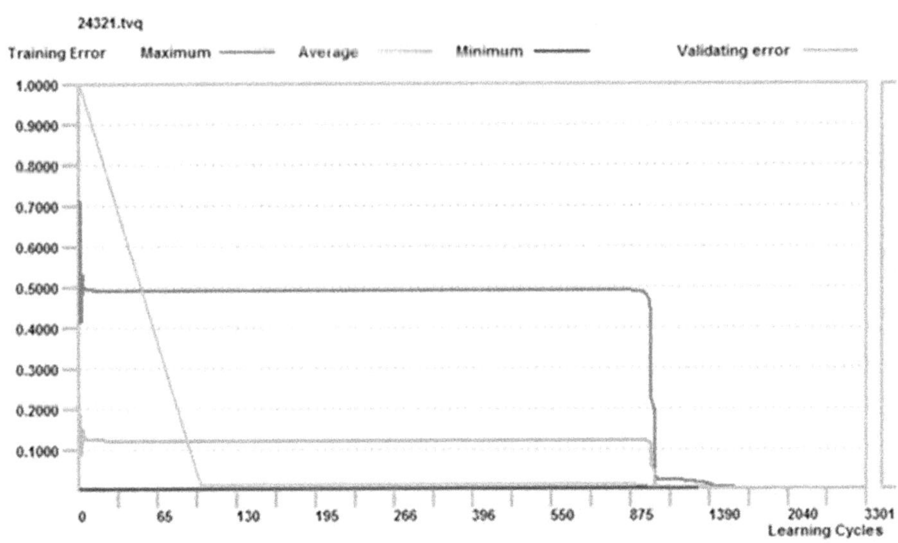

*Figure 14. Learning progress for three different ANN structure*

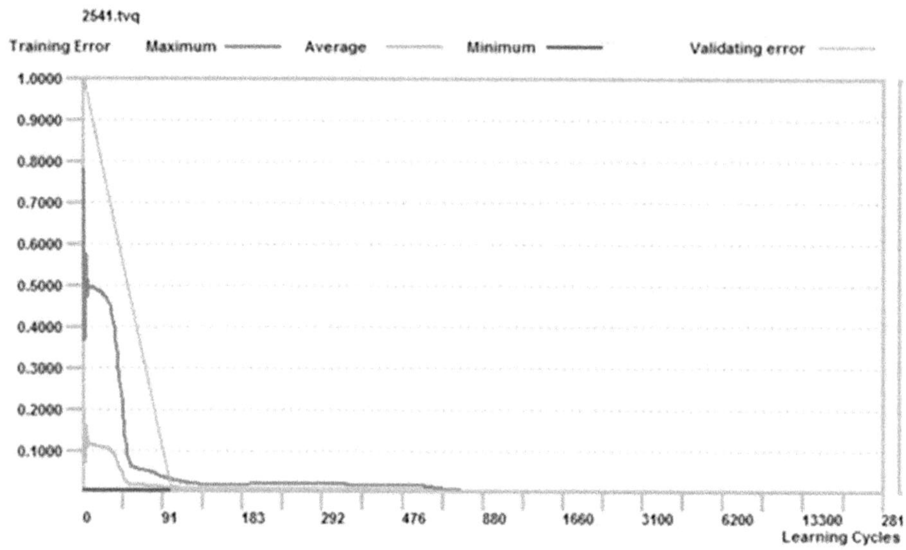


*Figure 15. Final chosen ANN structure (2*5*4*1)*

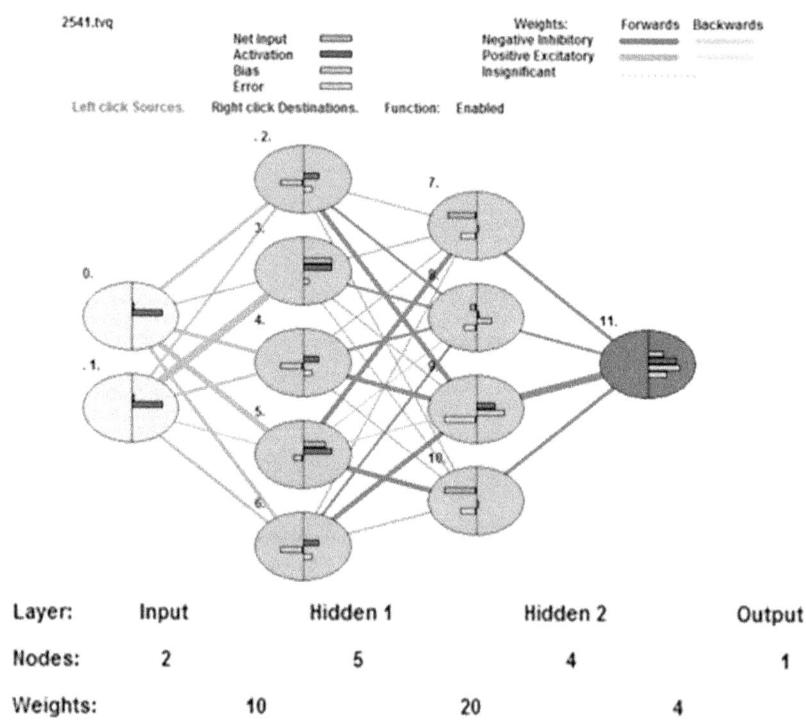

*Figure 16. The relation between the experimental data and the prediction model*

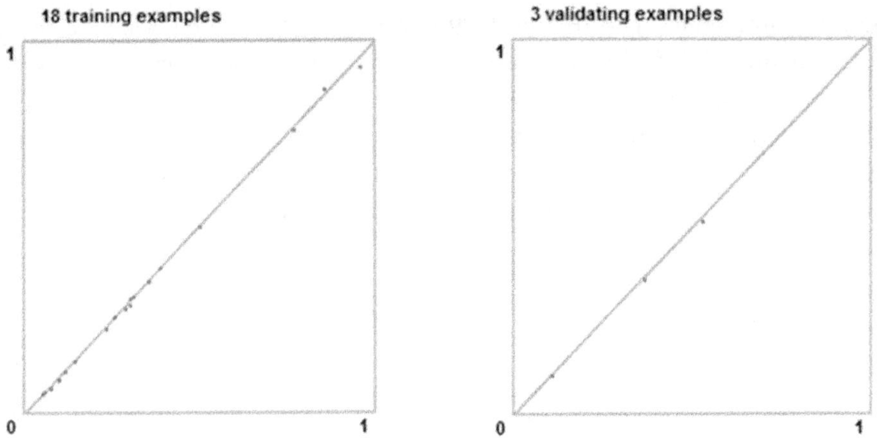


*Table 13. The differences between the actual measured and prediction data*

| Experiment No. | Actual value (Ra) in µmm | predicted value (Ra) in µmm | Error |
|---|---|---|---|
| 4 | 8.8 | 8.7039 | 0.0961 |
| 8 | 11.5 | 11.6967 | 0.1667 |
| 14 | 25.7 | 25.602 | 0.128 |
| 17 | 34 | 32.1928 | 1.8072 |
| 22 | 32.8 | 31.3344 | 1.4656 |
| 24 | 46.7 | 45.3149 | 1.3851 |

*Table 14. The sensitivity of the layer thickness and the building orientation*

| Input | Change from | to | sensitivity | Relative sensitivity |
|---|---|---|---|---|
| Layer thickness (mm) | 0.1 | 0.4 | 0.466 | |
| Orientation angle (deg.) | 0 | 80 | 0.246 | |

## Validation Process for Designed Structure of ANN

The minimum value of MSE is 1.2309 that is why the chosen structure in this implementation is 2*5*4*1. Figure 11 shows the network structure from EasyNN software version (17.0e – 2014). Figure 12 show the fitting of the model in the training samples. A comparison between the prediction of ANN model and the real measurement values for validation is presented in Table 13. The sensitivity of the layer thickness and the building orientation is shown in Table 14.

## CONCLUSION

In this chapter, regression models were Presented to predict the surface roughness through the building orientation. In addition, ANN was constructed to predict the surface roughness through the building orientation and the layer thickness together. Through this proposed analytical, practical study and the results, it can be concluded that:

- Layer thickness parameter is more effective than the building orientation parameter on the produced surface roughness.
- The experimental results show that horizontal and vertical surfaces usually have better surface quality that is why the part should be oriented to make surfaces are horizontal or vertical as possible.
- The early prediction of surface roughness, allow the manufacturer to make the better decision to fabricate the part either CNC or AM machine to satisfy the product requirements.

All the conclusions are "valid" for the given composition, processing parameters and test conditions.

This work can be extended in the future by doing some optimization work. Two process parameters (layer thickness and building orientation) can be studied to chive the minimum surface roughness.

## REFERENCES


Ahmad, M. N., Mohd, H. A. R., Maidin, N. A., Osman, M. H., Wahid, M. K., Hussin, M. S. F., & Nur, A. A. A. (2019). Optimization on Surface Roughness of Fused Deposition Modelling (FDM) 3D Printed Parts Using Taguchi Approach. *Symposium on Intelligent Manufacturing and Mechatronics SympoSIMM: Intelligent Manufacturing and Mechatronics*, 230-243.

Ahn, D., Kweon, J.-H., Choi, J.-H., & Lee, S.-H. (2011). Relation between surface roughness and overlap interval in Fused Deposition Modeling. *Advanced Materials Research, 264-265*, 1625–1630. doi:10.4028/www.scientific.net/AMR.264-265.1625

Boschetto, Giordano, & Veniali. (2013). Surface roughness prediction in fused deposition modelling by neural networks. *The International Journal of Advanced Manufacturing Technology*.

Gajdos, Spišák, Kaščák, & Krasinskyi. (2015). Surface Finish Techniques for FDM Parts. *Materials Science Forum, 818*, 45-48.

Hamoud, M. (2018a). Enhancement of Sliced Layer Contour for Reconstructing A 3D Model Using Solid Freeform Fabrication. *18th International Conference on Applied Mechanics and Mechanical Engineering-AMME-18*, 196-208. 10.21608/amme.2018.35012

Hamoud, M. (2018b). Surface Roughness Justification in Additive Manufacturing. *18th International Conference on Applied Mechanics and Mechanical Engineering-AMME-18*, 185-195. 10.21608/amme.2018.35011

Kandananond, K. (2020). Optimization of fused filament fabrication system by response surface method. *International Journal of Metrology and Quality Engineering, 11*, 4. doi:10.1051/ijmqe/2020002

Molero, Fernández, Rodríguez-Alabanda, Guerrero-Vaca, & Romero. (2020). Use of Data Mining Techniques for the Prediction of Surface Roughness of Printed Parts in Polylactic Acid (PLA) by Fused Deposition Modeling (FDM): A Practical Application in Frame Glasses Manufacturing. *Polymers, 2*(12), 840. doi:10.3390/polym12040840

Pérez, M., Medina-Sánchez, G., García-Collado, A., Gupta, M., & Carou, D. (2018). Surface Quality Enhancement of Fused Deposition Modeling (FDM) Printed Samples Based on the Selection of Critical Printing Parameters. *Materials (Basel), 11*(8), 1382. doi:10.3390/ma11081382 PMID:30096826

Sobhi, A., Hamoud, M., & Barakat, A., (2019). Optimal Building Orientation Based on Minimum Volumetric Error using a New Direct Slicing Algorithm. *International Journal of Science and Engineering Investigations, 8*(86).

Vahabli, E., & Rahmati, S. (2017). Improvement of FDM parts' surface quality using optimized neural networks medical case studies. *Rapid Prototyping Journal, 23*(4), 825–842. Advance online publication. doi:10.1108/RPJ-06-2015-0075

# Chapter 4
# Study and Application of Machine Learning Methods in Modern Additive Manufacturing Processes

**Ranjit Barua**
https://orcid.org/0000-0003-2236-3876
*CHST, Indian Institute of Engineering Science and Technology, Shibpur, India*

**Sudipto Datta**
*Indian Institute of Engineering Science and Technology, Shibpur, India*

**Pallab Datta**
*National Institute of Pharmaceutical Education and Research, Kolkata, India*

**Amit Roychowdhury**
*Indian Institute of Engineering Science and Technology, Shibpur, India*

## ABSTRACT

*Additive manufacturing (AM) make simpler the manufacturing of difficult geometric structures. Its possibility has quickly prolonged from the manufacture of pre-fabrication conception replicas to the making of finish practice portions driving the essential for superior part feature guarantee in the additively fabricated products. Machine learning (ML) is one of the encouraging methods that can be practiced to succeed in this aim. A modern study in this arena contains the procedure of managed and unconfirmed ML algorithms for excellent control and forecast of mechanical characteristics of AM products. This chapter describes the development of applying machine learning (ML) to numerous aspects of the additive manufacturing whole chain, counting model design, and quality evaluation. Present challenges in applying machine learning (ML) to additive manufacturing and possible solutions for these problems are then defined. Upcoming trends are planned in order to deliver a general discussion of this additive manufacturing area.*

DOI: 10.4018/978-1-7998-8516-0.ch004

## INTRODUCTION

Additive Manufacturing (AM), usually stated to as three dimensional printing or 3DP, is an advanced manufacturing technique that has the promising of disrupting the engineering industry on a balance not observed from the time when the manufacturing revolution (Barua et al.,2019). Paralleled to traditional manufacturing techniques, it has the benefits of manufacturing complicated parts with difficult geometric structure, exclusive microstructures and characteristics, in addition to concentrated lead period and budget. Consequently, in current existence time, additive manufacturing has involved an excessive arrangement of investigation awareness together industrial uses and academic research widespread. Additive manufacturing methods can be approximately categorized into seven classifications (Lee et al., 2001). Figure1 shows the different types of additive manufacturing process. The additive manufacturing methods connecting Machine Learning (ML) in this chapter generally classified into three groups, i.e. directed energy deposition (DED), powder bed fusion (PBF), and material extrusion process. Even though ML has also been used in additional additive manufacturing methods for example stereo-lithography (Wang et al., 2018) and materials jetting (Yuan et al., 2017). In powder bed fusion (PBF) process, an electron beam or laser is applied in place of the energy source to carefully melt powder bed which is homogeneously extent by repainting layer by layer (Barua et al., 2019). In the directed energy deposition (DED) method, a concentrated laser beam liquefies the incessant powder stream or wire which are deposited from nozzle obsessed by the melt pool so as to construct desire objects (Rahman et al., 2019). Fused deposition modelling (FDM) is an example of material extrusion procedure. The filament is melted by extruder heater and deposited layer by layer. Machine learning is an artificial intelligence (AI) method that permits a machine or system to study from data spontaneously and make conclusions or estimates without being plainly encoded (Datta et al., 2019). In the study, machine learning is in advance acceptance in health diagnostics (Ning et al.,2015), autonomous driving (Bruijne et al., 2016), smart manufacturing (Kourou et al., 2015), natural language processing (Datta et al., 2019) (Akerfeldt et al., 2016), object recognition (Liang et al., 2015) (LeCun et al., 2015), and material possessions expectation (Ward et al., 2016) (Pilania et al., 2013). ML algorithms are typically considered as supervised, unsupervised and reinforcement learning. Supervised learning allows a computer programme to study from a set of considered data in the teaching set so that it can classify unlabeled data from a trial set with the maximum probable exactness (Shi et al., 2016). The datasets can be in a diversity of forms counting forms of audio clips (O'Shea et al., 2016), images (Lempitsky et al., 2010), or typescript (Tong et al., 2001). There is a main function known as cost function, which analyses the mistake between the expected output standards and the real output standards. In the preparation method, the limitations (or weights) between neurons in neighboring layers are reorganized so as to decrease the cost function after every iteration (or epoch) (Daelemans dt al., 2003). In the testing procedure, the formerly hidden fresh data, specifically test set, is presented to deliver an impartial calculation of the simulation's accurateness. Unsupervised learning concludes from unlabeled data (Weber et al., 2000). It is a data-driven machine learning method which can expose unseen configurations or cluster comparable data collected in an assumed haphazard dataset (Alabi et al., 2018). Unsupervised learning is extensively applied in irregularity revealing (Omar et al., 2013), market dissection, and references systems (Tanev et al., 2007).

*Figure 1. Additive manufacturing process*

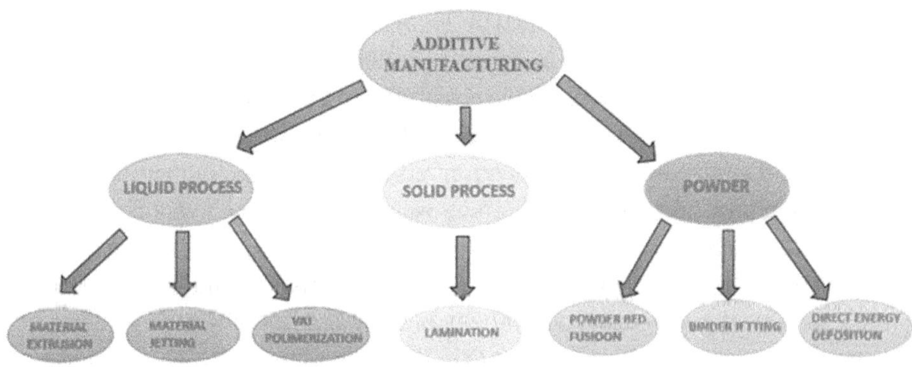


## APPLICATION OF MACHINE LEARNING IN ADDITIVE MANUFACTURING PROCESS

Usually, process or technique factor improvement and elevation are applied by simulation procedures or DOE (design of experiment) to additively fabrication new things. However, the DOE method generally contains trial-and-error, which is slow and expensive, mainly for metal additive manufacturing (Singh et al., 2012) (Mozaffar et al., 2018). Figure 2 shows the machine learning application in additive manufacturing process. The physical based model can expose the essential appliance for the development of explicit features throughout treating, for example micro-structure, melt pool geometry etc. (Tang et al., 2018). However, macro-scale models, for example finite element modelling, may undergo from divergences with investigational outcomes as a result of the easy expectations (Mozaffar et al., 2018). The progressively new refined methods, for example CFD (Computational Fluid Dynamics), generally application on particular trails (Tang et al., 2018) or a

*Figure 2. Machine learning using in additive manufacturing*

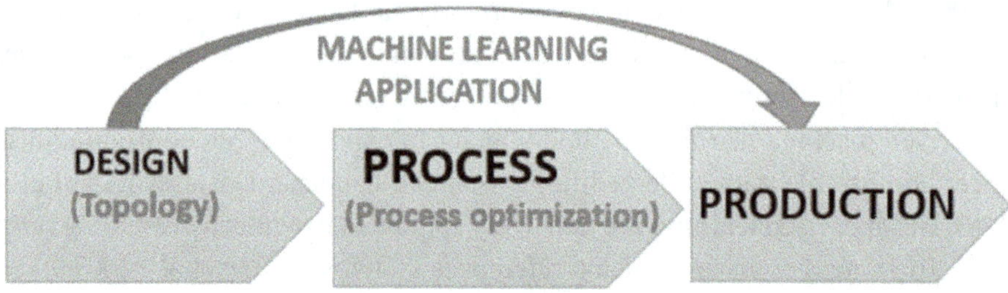


nominal amount of trails (Yan et al., 2018). This creates it stimulating to calculate the perfunctory characteristics of the amounts at a variety or macro range. Consequently, numerous scientists have discovered the possibility of introducing machine learning methods to explain the aforesaid tasks in procedure optimization of especially metal additive manufacturing process. It is observed that within the numerous additive manufacturing processes, machine learning was mostly applied to connection with their important procedure factors to the excellence displays at two different levels that is macroscale level (namely mechanical characteristics) and mesoscale level (namely melt pool geometries and porosity or comparative density). Furthermore, several scientists used machine learning to build progression plans, which could help as an outstanding conception tool to classify the practice frames. In case of the melt pool morphology, for example stability, equality and geometry, can mostly effect the finishing product feature. The mesoscale, single paths work as the important structure blocks of high-energy additive manufacturing procedures. Hence, MLP or multi-layer perceptron was applied to calculate mainly the height, depth and width) for wire-based (Xiong et al., 2014) and powder-based (Caiazzo et al., 2018). Directed energy deposition procedures founded on partial investigational datasets.

The MPG (Melt Pool Geometries) were later thoroughly connected to the procedure constraints. This suggests that a definite MPG is attainable by regulatory the procedure factors in an opposite mode. One more critical apprehension at mesoscale is the permeability of additive manufacturing constructed products. In case of metal additive manufacturing, reaching full concentration is the main target, equally the permeability considerably distresses the perfunctory performance of products, particularly fatigue characteristics (Lewandowski et al., 2016). The macro-scale possessions of additive manufacturing products can also be investigated by the machine learning method. ANFIS (Adaptive-network-based fuzzy inference system) generally can only perform relative facts. From this time it is good for the calculation of fatigue characteristics as a result of numerous uncertainties intricate in the fatigue procedure. The material extrusion procedure, the motivation of investigation is macro-scale mechanical characteristics. In fused deposition molding procedure, the procedure parameters that were widely examined contain the thickness of layer, raster angle and temperature. In this procedure, the best prevalent machine learning method is multi-layer perceptron. A correctly trained multi-layer perceptron displays advantage in apprehending the non-linear correlation of the structure for data appropriating and valuation abilities. Therefore, it was massively working to forecast creep and recovery properties (Mohamed et al., 2016), workable characteristics, wear rate (Scime et al., 2018), compressive strength (Sood et al., 2012), and dynamic modulus of elasticity (Ye et al., 2018) of Acrylonitrile Butadiene Styrene (ABS) and Polylactic Acid (PLA) materials. Also, Jiang et al., 2019, applied multi-layer perceptron to calculate the determined printable. Presently, the additive manufacturing procedure still undergoes from numerous processing-related faults for example distortion, delamination, rough surface, lack of fusion, cracks, porosity, process instability, and foreign inclusions. These flaws generally initiate from the layer-by-layer material deposition procedure. Certain may circulate from one layer to the succeeding layers, producing the complete shape to fail. Therefore, in-procedure observing acts an important character. Scientists have committed remarkable determinations to observing flaws throughout the printing process, which includes a sequences of monotonous phases and less fidelity. By the way, machine learning offers an original method for undertaking the experiments. Acoustic signals applied for acoustic-based observing mostly create from plasma in laser melting processing and perceptible stepper motor in fused deposition modelling process, correspondingly. The optical-based observing, the responses are mostly trail and scatter initials, strength, and melt pool outlines; in addition to layer by layer surface images taken by a variability of cameras. One motivating investigation is that the real-time CT (computed tomography) examination can become obtainable by including ex-situ computed tomography scanning into in-process observing with the assistance of machine learning process. Optical-based methods are the most generally accepted in-process observing approaches in the additive manufacturing area. Different apparatus like infrared (IR) thermal cameras and high-speed or digital cameras are generally applied to detention the optical pointers. In some circumstances, pyrometers and photodiodes can deliver additional evidence. Typically, signatures of the shape and sizes, scatter, top construct surface images, and temperature outlines of melt pools are composed as the efforts layer by layer. Making an allowance for the high feature sizes of the images, machine learning is increasing the acceptance in in-process observing of additive manufacturing procedures. Figure 3 shows the Multi-Layer Perceptron neural network. Thus, the machine learning based intelligent observing could deliver a superior clarification. In recent times, acoustic signals accumulated from the plasma that is produced at powder bed surface have been applied for in-process observing in powder bed fusion techniques. The overheating or under heating or of metal powder may alteration the surface temperature of slices, so altering the density of plasma. The distinction of plasma density composed with the variation of atmospheric compression in the surrounded chamber impacts the acoustic strength. As related to the conventional machine learning techniques which include many successive phases, for example feature extraction, data processing etc. deep belief networks can make things easier and faster the processing by discriminative fine-tuning and reproductive pre-training. In the fused deposition modelling techniques, the audio waves released from the extruder and printed products can be identified by sensors. The K-means clustering was applied to distinguish usual and unsuccessful printing procedure, as the alteration in audio feature designs reproduces the incidence of irregularities. Also, they attained the similar objective by using the semi-Markov model with decreased feature sizes and earlier data dispensation (Wu et al., 2017). In directed energy deposition techniques, thermal phases of melt pools were managed by SOM (self-organizing map) to classify irregular melt pool and expect permeability with the assistance of ex-situ computed tomography scan data (Jafri-Marandi et al., 2019). This structure could help as a real-time computed tomography scanner for laser based additive manufacturing techniques.

## ADDITIVE MANUFACTURING DESIGN USING MACHINE LEARNING

Because of the boundless design of freedom in additive manufacturing (DfAM), this DfAM is expressively not like from the principles of designs which are mainly practiced in the normal manufacturing process. The applications of the machine learning in DfAM will be clarified into two parts, i.e. material design and topological design.

### Application of Machine Learning in Topology Design

Inappropriately, the related research work of using machine learning for topological design for the additive manufacturing is still inadequate. The mixed machine learning method for additive manufacturing design feature endorsement at the time of conceptual design stage by support vector machine (SVM) and hierarchical clustering (unsupervised machine learning) was proposed by Yao et al. 2017. Though, this research only replaces the huge structures in the main design along with light structures which are imported through a database that can not include any topology optimization procedures. Topology optimization is an orderly process that produces structures using material distribution optimization inside a provided

*Figure 3. Multi-layer perceptron network*

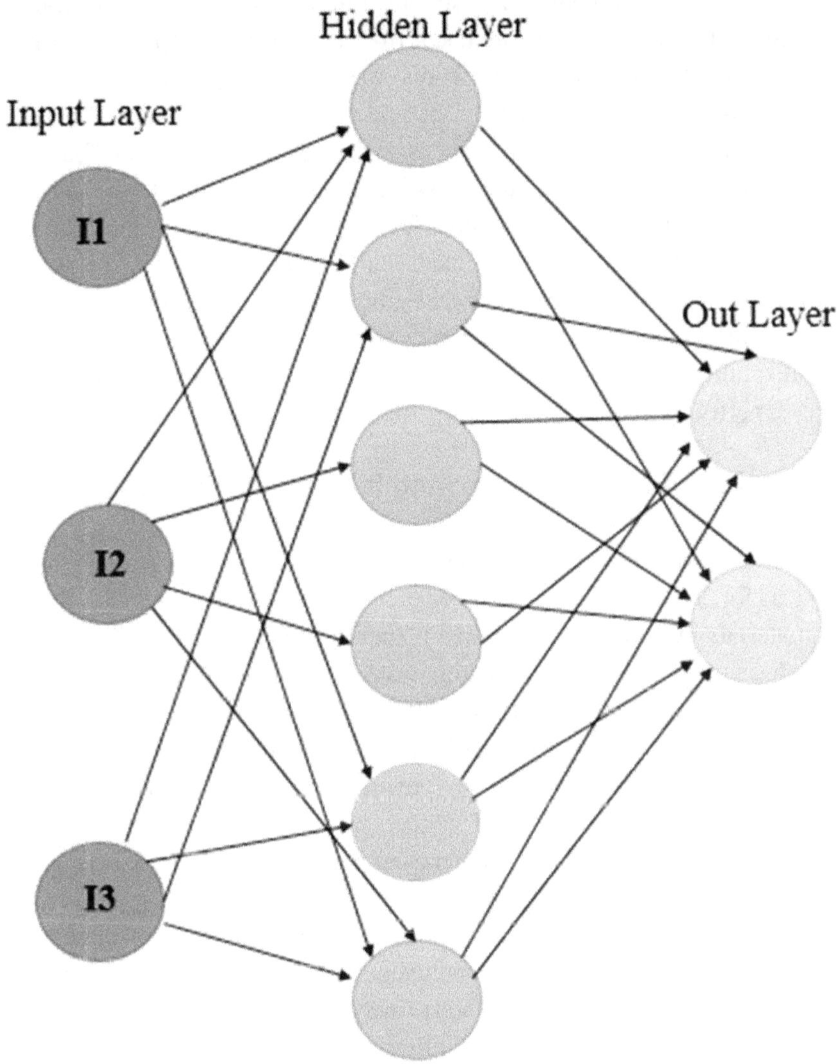


space subject to suitable constraints and loads (Banga et al., 2018). Ideally, normal topology optimization process needs a huge repetitions of prototyping and design and thus computationally huge demanding mainly for complex structures and large-scale. Machine learning models, mainly for huge deep neural networking, at the time of training stage also faces this challenges (Rawat et al., 2018). Though, once the machine learning structures are highly trained then can produce desirable designs fast without having to again start from the scratch, which finally allows the machine learning- centric method for providing opposite alternative against the topology optimization method. For solving a mechanical problem, for training the intermediate topologies which are taken from conventional topology optimization process, convolutional neural networks was used. For example the topology optimization solver was stopped at a middle phage only after some repetitions for predicting the best optimized structure (Sosnovik et al., 2019). It was found out that the qualified convolutional neural networks model can estimate the net topology optimized structure nearly twenty times faster than the standard simplified isotropic material with penalization (SIMP) method (Yu et al., 2018). This given proposal pipeline can also be used for solving heating conduction problem, and it outdone SIMP results in binary and speed accuracy with the thresholding. It shows the robust generalizability of the convolutional neural networks structure deprived of trusting on the expertise of the problem type.

## Application of Machine Learning in Material Design

Metamaterials are one kind of material produced by material scientists and engineers for decades which can exceed their basic bilk materials. Though, creating metamaterials by Edisonian approach manually is very difficult and challenging. This is because of possible combinations astronomical number. With the help of the machine learning process the discovery method of the metamaterials can be highly accelerated (Gu et al., 2015). The present advancement of the machine learning helped the engineers and the material scientists to jump from expecting material properties (Liu et al., 2015) (Silver et al., 2017) for constructing new metamaterials (Ma et al., 2018). Furthermore, additive manufacturing process can materialize designs which were impractical to construct, as shown my many scientists works (Li et al., 2017) (Vyatskikh et al., 2018). The possible for the interaction of state-of-the-art machine learning in additive manufacturing techniques and materials design leftovers fairly untouched. A totally automated method to invent best structures of the metamaterials was developed by Chen et al. 2018; afterwards it was verified by selective laser sintering (SLS) process along with PEBA2301 elastic material. It is intended that given the wanted elastic material properties, i.e. poisson's ratio, shear modulus and Young's modulus, the arrangement can produce modified microstructure that equals the description by means of Machine Learning (Pham et al., 2019). A deep learning-based procedure monitoring technique was analyzed for DED (directed energy deposition) in additive manufacturing (Wang et al., 2020). The thermal images composed at the time experiment are practiced to classify the process state, and a deep convolutional neural network model is suggested to construct an end-to-end condition monitoring framework. Investigates on an actually focused energy deposition dataset in additive manufacturing are performed for authentication. The outcomes suggest the planned technique offers a hopeful methodology in-process monitoring based on the industrial images. Convolutional neural networks was then given for training the database where the mechanical properties were measured by using FEM which produced novel microstructural formations of the metamaterial composite which was two times harder and forty times tougher.

## APPLICATION OF MACHINE LEARNING IN DATA SECURITY

In many fields intelligent property (IP) protection is takes as a high importance. In simple words, digital marketing contains mainly two important modules i.e. a) physical domain and b) cyber domain. Because of the growing dependency on digitalization and IoT (Internet-of-Things) (Li et al., 2015), many security cases, for example, malware attack (McIntosh et al., 2019), denial of service (DoS), and illegal access (Sun et al., 2018), data breach (Shaw et al., 2009), zero-day attack (Alazab et al., 2010), social networking or phishing (Gupta et al., 2017), etc. have developed progressively in latest years. According to Papastergiou et al., 2019, the national security of a country is influenced by the government, business, and commoners who use different social networking tools and apps that are extremely secure, and the ability to identify and remove such cyber-bullies in an appropriate manner. Day-by-day, the need for these similar technologies is gradually increasing, which is shown in Figure 5, depends upon the information of the past 12 months which is gathered from Google Trends (Google trends, 2021). Though, data cracking is normally happens by cyber domain, but sometimes it can also occur through physical domain which is also recognized as side channels, because additive manufacturing can produce many signals when producing three-dimensional products. IP spying can take benefit of Machine Learning methods to procedure the produced signs to rebuild computer-aided design data incidentally. Till now, using machine learning for three-dimensional object construction from the size channels is proved to be possible generally through gathering acoustic signals at the time of printing. The acoustic signals of the stepper motors of a fused deposition modelling can be gathered by using a mic. This signal can incidentally reproduce G-code, which allows the information to leak, information like speed of nozzle, material extrusion amount, and temperature and axis movement for the fused deposition modelling process. According to Faruque et al. 2016, the acoustic data extracted features can be used for training machine learning algorithms for reconstructing a key model having length prediction accuracy 18% and axis prediction accuracy 78%. In a more hard-to-detect IP stealing situation, the spying can even place his mobile phone near the apparatus to arrest the acoustic data. Hojjati et al. 2016, have effectively taken benefit of this information to rebuild an airplane model inside $1^0$ of error in angle and 1 mm error in length.

*Figure 4. Directed energy deposition (DED)*

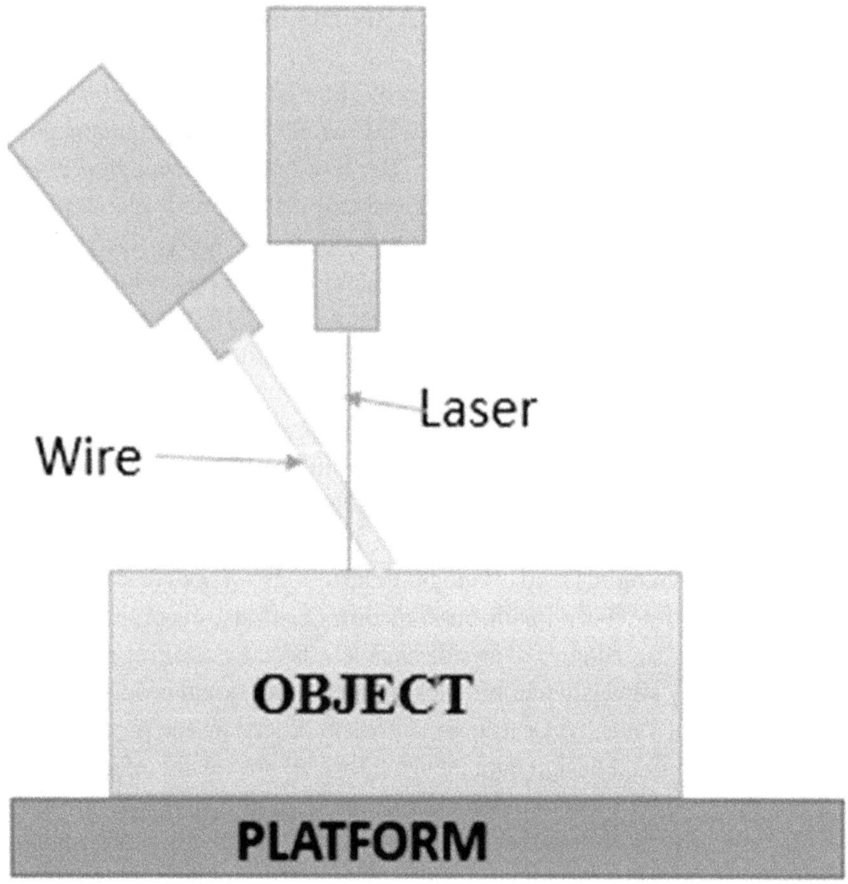


## APPLICATION OF MACHINE LEARNING IN QUALITY CONTROL

One complex issue that delays the additive manufacturing products authorization is the discrepancy of manufactured goods superiority from apparatus to apparatus of the same procedure, otherwise from build to build of the similar apparatus. This discrepancy can lead to variations in relative density, geometrical accuracy,

*Figure 5. Worldwide popularity trends of artificial intelligence, machine learning, cyber-security, and data science (From 1st Jan, 2020 – 13th June, 2021)*

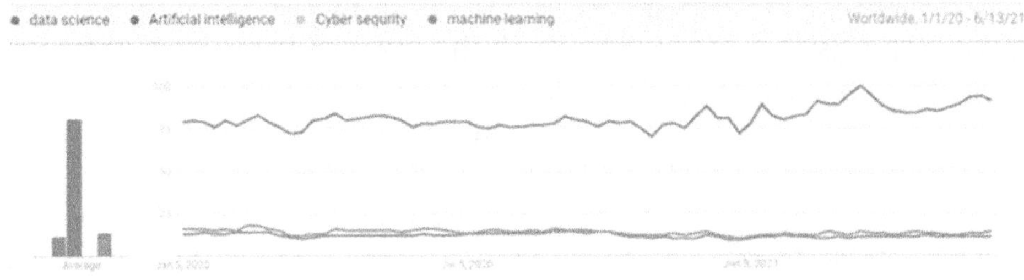


mechanical properties and process immovability. That is why costly research works have tried to provide machine learning processes to get quality control of the additive manufacturing parts. The errors in the geometry can be minimized by implementing process control, modifying the original computer-aided design and namely rescaling the total process. Before the production (Baturynska et al., 2018) of the parts the scaling ration can be expected by convolutional neural networks or multi-layer perceptron for regulating the total size of the parts. By machine learning algorithms the shape reliant on symmetrical deviances because of thermal stress can be regulated for making computer-aided design file geometrical modifications. More exactly, multi-layer perceptron was applied to recompense for geometrical distortion to counter the thermal effects causing from Selective Laser Modelling processing, as confirmed by Chowdhury et al. 2018. FEM imitation data were skilled to forecast the distorted sites in order to adjust the novel computer-aided design shape. Noriega et al. 2013 applied a same method in fused deposition modelling printing, where simulated data was not trained only the experimental data were changed. To attain procedure regulator, self-organizing map can link exact kinds of geometric nonconformities to sure procedure situations (Khanzadeh et al., 2018). This process can highly decrease the three-dimensional point cloud amount which is needed when entering geometrical accuracy of the additive manufacturing parts by the help of laser scanner, as compared to various mainstream managed machine learning methods (Tang et al., 2017). Furthermore, by process parameters controlling for directed energy deposition, the single tracks shape can be changed for decreasing the geometric errors in the macro-scale level (Caiazzo et al., 2018) (Xiong et al., 2014). Figure 6 shows the uses of machine learning to analyze several parameters during the additive manufacturing process. Surface images which are taken in the each produced layer after the exposure of the laser for training the machine learning algorithm is done in the powder bed fusion process for warped parts early detection before the powder coating was done (Grasso et al., 2017). As discussed early for improving the process stability, mechanical performance and the relative build quality of additive manufacturing parts, in-process monitoring is hired by presenting various cameras and sensors as deliberated before. The emitted signals are mainly acoustic signals and visual signals which are processed and collected for training different machine learning algorithms for monitoring the process of the printing. In additive manufacturing, machine learning can be used to mechanically identify the status of the printing (Rao et al., 2015) (Uhlmann et al., 2017), tensile property prediction (Okaro et al., 2018), surface roughness prediction (Li et al., 2019), failure modes (Wu et al., 2019) (He et al., 2018), porosity detection (Khanzadeh et al., 2019) and melting condition (Li et al., 2019) (Renken et al., 2017).

*Figure 6. Several parameters estimated by machine learning*

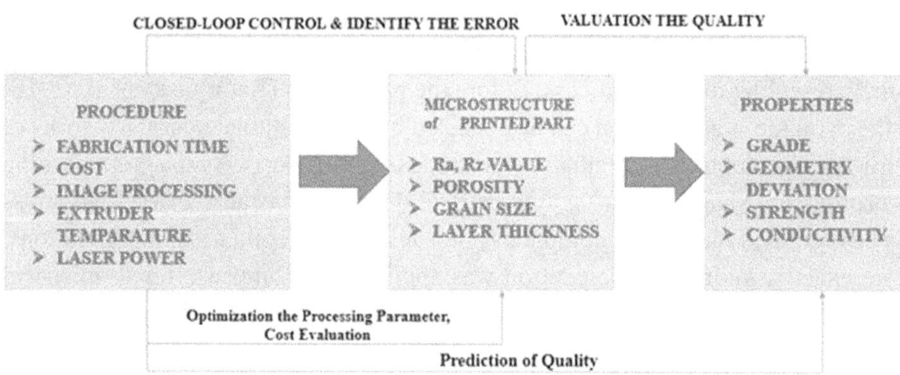


## CONCLUSION

3D printing also known as additive manufacturing is expanding at a very fast rate in the manufacturing industry and gained a huge importance from many domains because of its ability to create parts having complex features. The dependence of the 3D printed products is the main focus of the scientists to realize additive manufacturing as a last production tool. In Additive manufacturing process also, machine learning is also being applied for modifying the total manufacturing and design algorithm mainly in the time of 4.0 industry. In this chapter we shall discuss total types of machine learning processes. After that we shall discuss the many aspects of the machine learning in many fields of the additive manufacturing processes like material tuning, cybersecurity, cloud service, in situ monitoring and process optimization. We shall also give light of the machine learning in the field of tissue engineering and biomedical engineering construction and building. The various challenges which are faced by the machine learning in additive manufacturing like standards for qualification, data acquisition techniques and computational cost will also be explained. From the authors point of view additive manufacturing process in situ monitoring will highly be useful from the object identification capability of the machine learning. As a huge data set is highly needed for machine learning, additive manufacturing data sharing will activate fast adaptation of the machine learning in the additive manufacturing. Criteria for the data shared are wanted to simplify easy data sharing. The using of Machine learning in Additive manufacturing will be more developed and hugely accept as a best data achievement techniques and more advanced computer chips for machine learning are being developed.

## ACKNOWLEDGMENT

The authors would like to thank IIEST-Shibpur, CHST lab, and thanks to Mrs. Nibedita Bardhan for language proof reading.

## REFERENCES

Akerfeldt, P., Antti, M.-L., & Pederson, R. (2016). Influence of microstructure on mechanical properties of laser metal wire-deposited Ti-6Al-4V. *Materials Science and Engineering A*, *674*, 428–437. doi:10.1016/j.msea.2016.07.038

Alabi, M. O., Nixon, K., & Botef, I. (2018). A survey on recent applications of machine learning with big data in additive manufacturing industry. *American Journal of Engineering and Applied Sciences*, *11*(3), 1114–1124. doi:10.3844/ajeassp.2018.1114.1124

Alazab, M., Venkatraman, S., Watters, P., & Alazab, M. (2010). *Zero-day malware detection based on supervised learning algorithms of API call signatures*. Academic Press.

Banga, Gehani, Bhilare, Patel, & Kara. (2018). *3D topology optimization using convolutional neural networks*. arXiv 1808.07440.

Barua, R., Datta, S., Datta, P., & Chowdhury, A. R. (2019). Scaffolds and Tissue Engineering Applications by 3D Bio-Printing Process: A New Approach. In K. Kumar & J. Davim (Eds.), *Design, Development, and Optimization of Bio-Mechatronic Engineering Products* (pp. 78–99). IGI Global. doi:10.4018/978-1-5225-8235-9.ch004

Barua, R., Datta, S., Roychowdhury, A., & Datta, P. (2019). Importance of 3D Printing Technology in Medical Fields. In Additive Manufacturing Technologies From an Optimization Perspective (pp. 21-40). IGI Global. doi:10.4018/978-1-5225-9167-2.ch002

Baturynska, I., Semeniuta, O., & Wang, K. (2018). Application of machine learning methods to improve dimensional accuracy in additive manufacturing. *International Workshop of Advanced Manufacturing and Automation*, 245–252.

Bruijne, M. (2016). Machine learning approaches in medical image analysis: From detection to diagnosis. *Medical Image Analysis*, *33*, 94–97. doi:10.1016/j.media.2016.06.032 PMID:27481324

Caiazzo, F., & Caggiano, A. (2018). Laser direct metal deposition of 2024 Al alloy: Trace geometry prediction via machine learning. *Materials (Basel)*, *11*(3), 444. doi:10.3390/ma11030444 PMID:29562682

Chen, D., & Zheng, X. (2018). Multi-material additive manufacturing of metamaterials with giant, tailorable negative Poisson's ratios. *Scientific Reports*, *8*(9139), 9139. doi:10.103841598-018-26980-7 PMID:29904093

Chowdhury, S., Mhapsekar, K., & Anand, S. (2018). Part build orientation optimization and neural network-based geometry compensation for additive manufacturing process. *Journal of Manufacturing Science and Engineering*, *140*(3), 031009. doi:10.1115/1.4038293

Daelemans, W., Hoste, V., Meulder, F. D., & Naudts, B. (2003). Combined optimization of feature selection and algorithm parameters in machine learning of language. *European Conference on Machine Learning*, 84–95. 10.1007/978-3-540-39857-8_10

Datta, S., Barua, R., & Das, J. (2019). *Application of Artificial Intelligence in Modern Healthcare System, Alginates - Recent Uses of This Natural Polymer, Leonel Pereira*. IntechOpen. Available from: https://www.intechopen.com/books/alginates-recent-uses-of-this-natural-polymer/application-of-artificial-intelligence-in-modern-healthcare-system doi:10.5772/intechopen.90454

Faruque, A., Abdullah, M., Chhetri, S. R., Canedo, A., & Wan, J. (2016). Acoustic side-channel attacks on additive manufacturing systems. *Proceedings of the 7th International Conference on Cyber-Physical Systems*, 1–10. 10.1109/ICCPS.2016.7479068

Gu, G. X., Chen, C.-T., Richmond, D. J., & Buehler, M. J. (2018). Bioinspired hierarchical composite design using machine learning: Simulation, additive manufacturing, and experiment. *Materials Horizons*, *5*(5), 939–945. doi:10.1039/C8MH00653A

Gupta, B. B., Tewari, A., Jain, A. K., & Agrawal, D. P. (2017). Fighting against phishing attacks: State of the art and future challenges. *Neural Computing & Applications*, *28*(12), 3629–3654. doi:10.100700521-016-2275-y

He, K., Yang, Z., Bai, Y., Long, J., & Li, C. (2018). Intelligent fault diagnosis of delta 3D printers using attitude sensors based on support vector machines. *Sensors (Basel)*, *18*(4), 1298. doi:10.339018041298 PMID:29690641

Hojjati, A., Adhikari, A., Struckmann, K., Chou, E., Nguyen, T. N. T., Madan, K., Winslett, M. S., Gunter, C. A., & King, W. P. (2016). Leave your phone at the door: Side channels that reveal factory floor secrets. *Proceedings of the 2016 ACM SIGSAC Conference on Computer and Communications Security*, 883–894. 10.1145/2976749.2978323

Jafari-Marandi, R., Khanzadeh, M., Tian, W., Smith, B., & Bian, L. (2019). From in-situ monitoring toward high-throughput process control: Cost-driven decision-making framework for laser-based additive manufacturing. *Int. J. Ind. Manuf. Syst. Eng.*, *51*, 29–41. doi:10.1016/j.jmsy.2019.02.005

Jiang, J., Hu, G., Li, X., Xu, X., Zheng, P., & Stringer, J. (2019). Analysis and prediction of printable bridge length in fused deposition modelling based on back propagation neural network. *Virtual and Physical Prototyping*, *14*(3), 253–266. doi:10.1080/17452759.2019.1576010

Kourou, K., Exarchos, T. P., Exarchos, K. P., Karamouzis, M. V., & Fotiadis, D. I. (2015). Machine learning applications in cancer prognosis and prediction. *Computational and Structural Biotechnology Journal*, *13*, 8–17. doi:10.1016/j.csbj.2014.11.005 PMID:25750696

LeCun, Y., Bengio, Y., & Hinton, G. (2015). Deep learning. *Nature*, *521*(7553), 436–444. doi:10.1038/nature14539 PMID:26017442

Lee, Park, Cho, Zhang, & Leu. (2001). A neural network approach to the modelling and analysis of stereolithography processes. *Proc. Instn. Mech. Engrs., 215*(Part B), 1719–1733.

Lempitsky, V., & Zisserman, A. (2010). Learning to count objects in images. *Advances in Neural Information Processing Systems*, 1324–1332.

Lewandowski, J. J., & Seifi, M. (2016). Metal additive manufacturing: A review of mechanical properties. *Annual Review of Materials Research*, *46*(1), 151–186. doi:10.1146/annurev-matsci-070115-032024

Li, S., Da Xu, L., & Zhao, S. (2015). The internet of things: A survey. *Information Systems Frontiers*, *17*(2), 243–259. doi:10.100710796-014-9492-7

Li, S., Hassanin, H., Attallah, M. M., Adkins, N. J., & Essa, K. (2016). The development of TiNi based negative Poisson's ratio structure using selective laser melting. *Acta Materialia*, *105*, 75–83. doi:10.1016/j.actamat.2015.12.017

Liang, M., & Hu, X. (2015). Recurrent convolutional neural network for object recognition. *Proceedings of the IEEE Conference on Computer Vision and Pattern Recognition*, 3367–3375. 10.1109/CVPR.2015.7298958

Liu, R., Kumar, A., Chen, Z., Agrawal, A., Sundararaghavan, V., & Choudhary, A. (2015). A predictive machine learning approach for microstructure optimization and materials design. *Scientific Reports*, *5*(11551), 11551. doi:10.1038rep11551 PMID:26100717

Ma, W., Cheng, F., & Liu, Y. (2018). Deep-learning-enabled on-demand design of chiral metamaterials. *ACS Nano*, *12*(6), 6326–6334. doi:10.1021/acsnano.8b03569 PMID:29856595

McIntosh, T., Jang-Jaccard, J., Watters, P., & Susnjak, T. (2019). The inadequacy of entropy-based ransomware detection. In *International conference on neural information processing*. New York: Springer.

Mohamed, O. A., Masood, S. H., & Bhowmik, J. L. (2017). Influence of processing parameters on creep and recovery behavior of FDM manufactured part using definitive screening design and ANN. *Rapid Prototyping Journal*, *23*(6), 998–1010. doi:10.1108/RPJ-12-2015-0198

Mozaffar, M., Paul, A., Al-Bahrani, R., Wolff, S., Choudhary, A., Agrawal, A., Ehmann, K., & Cao, J. (2018). Data-driven prediction of the high-dimensional thermal history in directed energy deposition processes via recurrent neural networks. *Manufacturing Letters*, *18*, 35–39. doi:10.1016/j.mfglet.2018.10.002

Ning, F., Cong, W., Qiu, J., Wei, J., & Wang, S. (2015). Additive manufacturing of carbon fiber reinforced thermoplastic composites using fused deposition modeling. *Composites. Part B, Engineering*, *80*, 369–378. doi:10.1016/j.compositesb.2015.06.013

O'Shea, T. J., Corgan, J., & Clancy, T. C. (2016). Convolutional radio modulation recognition networks. *International Conference on Engineering Applications of Neural Networks*, 213–226. 10.1007/978-3-319-44188-7_16

Okaro, Jayasinghe, Sutcliffe, Black, Paoletti, & Green. (2018). *Automatic Fault Detection for Selective Laser Melting Using Semi-Supervised Machine Learning.* Academic Press.

Omar, S., Ngadi, A., & Jebur, H. H. (2013). Machine learning techniques for anomaly detection: An overview. *Int. J. Comput. Appl. Technol. Res.*, *79*(2), 33–41. doi:10.5120/13715-1478

Papastergiou, S., Mouratidis, H., & Kalogeraki, E.-M. (2019). Cyber security incident handling, warning and response system for the european critical information infrastructures (cybersane). In *International Conference on Engineering Applications of Neural Networks* (pp. 476–487). New York: Springer. 10.1007/978-3-030-20257-6_41

Pham, M. S., Liu, C., Todd, I., & Lertthanasarn, J. (2019). Damage-tolerant architected materials inspired by crystal microstructure. *Nature*, *565*(7739), 305–311. doi:10.103841586-018-0850-3 PMID:30651615

Pilania, G., Wang, C., Jiang, X., Rajasekaran, S., & Ramprasad, R. (2013). Accelerating materials property predictions using machine learning. *Scientific Reports*, *3*(1), 2810. doi:10.1038rep02810 PMID:24077117

Rahman, N. U., Capuano, L., Cabeza, S., Feinaeugle, M., Garcia-Junceda, A., de Rooij, M. B., Mathews, D. T. A., Walmag, G., Gibson, I., & Römer, G. R. B. E. (2019). Directed energy deposition and characterization of high-carbon high speed steels. *Additive Manufacturing*, *30*(100838). doi:10.1016/j.addma.2019.100838

Rao, P. K. P., Liu, J. P., Roberson, D., Kong, Z. J., & Williams, C. (2015). Online real-time quality monitoring in additive manufacturing processes using heterogeneous sensors. *Journal of Manufacturing Science and Engineering*, *137*(6), 061007. doi:10.1115/1.4029823

Rawat & Shen. (2018). *A novel topology design approach using an integrated deep learning network architecture.* arXiv 1808.02334.

Renken, V., Albinger, S., Goch, G., Neef, A., & Emmelmann, C. (2017). Development of an adaptive, self-learning control concept for an additive manufacturing process. *CIRP Journal of Manufacturing Science and Technology*, *19*, 57–61. doi:10.1016/j.cirpj.2017.05.002

Scime, L., & Beuth, J. (2018). Anomaly detection and classification in a laser powder bed additive manufacturing process using a trained computer vision algorithm. *Additive Manufacturing*, *19*, 114–126. doi:10.1016/j.addma.2017.11.009

Shaw, A. (2009). Data breach: From notification to prevention using pci dss. *Colum Soc Probs.*, *43*, 517.

Shi, B., Bai, X., & Yao, C. (2016). An end-to-end trainable neural network for image-based sequence recognition and its application to scene text recognition. *IEEE Transactions on Pattern Analysis and Machine Intelligence*, *39*(11), 2298–2304. doi:10.1109/TPAMI.2016.2646371 PMID:28055850

Silver, D., Schrittwieser, J., Simonyan, K., Antonoglou, I., Huang, A., Guez, A., Hubert, T., Baker, L., Lai, M., Bolton, A., Chen, Y., Lillicrap, T., Hui, F., Sifre, L., Driessche, G. V. D., Graepel, T., & Hassabis, D. (2017). Mastering the game of go without human knowledge. *Nature*, *550*(7676), 354–359. doi:10.1038/nature24270 PMID:29052630

Singh, A., Cooper, D., Blundell, N., Gibbons, G., & Pratihar, D. (2012). Modelling of direct metal laser sintering of EOS DM20 bronze using neural networks and genetic algorithms. *Proceedings of the 37th International MATADOR Conference.*

Sood, A. K., Ohdar, R. K., & Mahapatra, S. S. (2012). Experimental investigation and empirical modelling of FDM process for compressive strength improvement. *Journal of Advanced Research*, *3*(1), 81–90. doi:10.1016/j.jare.2011.05.001

Sosnovik, I., & Oseledets, I. (2019). Neural networks for topology optimization. *Russ. J. Number Anal. M.*, *34*(4), 215–223. doi:10.1515/rnam-2019-0018

Sun, N., Zhang, J., Rimba, P., Gao, S., Zhang, L. Y., & Xiang, Y. (2018). Data-driven cybersecurity incident prediction: A survey. *IEEE Communications Surveys and Tutorials*, *21*(2), 1744–1772. doi:10.1109/COMST.2018.2885561

Tanev, H. (2007). Unsupervised learning of social networks from a multiple-source news corpus. *Proceedings of the Workshop Multi-Source Multilingual Information Extraction and Summarization*, 33–40.

Tang, C., Tan, J., & Wong, C. (2018). A numerical investigation on the physical mechanisms of single track defects in selective laser melting. *Int. J. Heat Mass Transf. – Theory Appl.*, *126*, 957–968.

Tong, S., & Koller, D. (2001). Support vector machine active learning with applications to text classification. *Journal of Machine Learning Research*, *2*, 45–66.

Uhlmann, E., Pontes, R. P., Laghmouchi, A., & Bergmann, A. (2017). Intelligent pattern recognition of a SLM machine process and sensor data. *Procedia CIRP*, *62*, 464–469. doi:10.1016/j.procir.2016.06.060

Vyatskikh, A., Delalande, S., Kudo, A., Zhang, X., Portela, C. M., & Greer, J. R. (2018). Additive manufacturing of 3D nano-architected metals. *Nature Communications*, *9*(1), 593. doi:10.103841467-018-03071-9 PMID:29426947

Wang, C., Tan, X. P., Tor, S. B., & Lim, C. S. (2020). Machine learning in additive manufacturing: State-of-the-art and perspectives. *Additive Manufacturing*, *36*, 101538. Advance online publication. doi:10.1016/j.addma.2020.101538

Wang, T., Kwok, T.-H., Zhou, C., & Vader, S. (2018). In-situ droplet inspection and closed-loop control system using machine learning for liquid metal jet printing. *Int. J. Ind. Manuf. Syst. Eng.*, *47*, 83–92. doi:10.1016/j.jmsy.2018.04.003

Ward, L., Agrawal, A., Choudhary, A., & Wolverton, C. (2016). A general-purpose machine learning framework for predicting properties of inorganic materials. *NPJ Comput. Mater.*, *2*, 16028.

Weber, M., Welling, M., & Perona, P. (2000). Unsupervised learning of models for recognition. *European Conference on Computer Vision*, 18–32.

Wu, H., Yu, Z., & Wang, Y. (2016). A new approach for online monitoring of additive manufacturing based on acoustic emission. *Proceedings of the ASME 2016 International Manufacturing Science and Engineering Conference*. 10.1115/MSEC2016-8551

Wu, M., Zhou, H., Lin, L. L., Silva, B., Song, Z., Cheung, J., & Moon, Y. (2017). Detecting attacks in Cyber Manufacturing systems: Additive manufacturing example. *MATEC Web of Conferences*, *108*, 06005. doi:10.1051/matecconf/201710806005

Xiong, J., Zhang, G., Hu, J., & Wu, L. (2014). Bead geometry prediction for robotic GMAW based rapid manufacturing through a neural network and a second-order regression analysis. *Journal of Intelligent Manufacturing*, *25*(1), 157–163. doi:10.100710845-012-0682-1

Yan, W., Qian, Y., Ge, W., Lin, S., Liu, W. K., Lin, F., & Wagner, G. J. (2018). Meso-scale modeling of multiple-layer fabrication process in selective electron beam melting: Interlayer/track voids formation. *Materials & Design*, *141*, 210–219. doi:10.1016/j.matdes.2017.12.031

Yao, X., Moon, S. K., & Bi, G. (2017). A hybrid machine learning approach for additive manufacturing design feature recommendation. *Rapid Prototyping Journal, 23*(6), 983–997. doi:10.1108/RPJ-03-2016-0041

Ye, D., Hong, G. S., Zhang, Y., Zhu, K., & Fuh, J. Y. H. (2018). Defect detection in selective laser melting technology by acoustic signals with deep belief networks. *International Journal of Advanced Manufacturing Technology, 96*(5-8), 2791–2801. doi:10.100700170-018-1728-0

Yu, Hur, & Jung. (2018). *Deep learning for topology optimization design*. arXiv 1801.05463.

Yuan, S., Shen, F., Bai, J., Chua, C. K., Wei, J., & Zhou, K. (2017). 3D soft auxetic lattice structures fabricated by selective laser sintering: TPU powder evaluation and process optimization. *Materials & Design, 120*, 317–327. doi:10.1016/j.matdes.2017.01.098

## KEY TERMS AND DEFINITIONS

**3DP:** Three-dimensional printing is also denoted to as AM or additive manufacturing process. In this additive manufacturing process, one makes a design of the desire item by CAD software, and the printer fabricates the desire product by depositing material layer by layer.

**Additive Manufacturing (AM):** Additive manufacturing (AM) is also known as ALM (additive layer manufacturing) process, which is the modern manufacturing fabrication name for three-dimensional printing (3DP), basically a computer controlled method that makes 3D objects layer by layer liquid material depositing.

**Directed Energy Deposition (DED):** Directed energy deposition (DED) is a three-dimensional printing process which practices a focused energy source, for example the electron beam, laser, or a plasma arc to liquefy a material which is instantaneously dropped from nozzle (Figure 4).

**Machine Learning:** Machine learning offers systems the facility to automatically study and recover from exercise without being openly programmed, it is also known as artificial intelligence (AI). Generally it emphases on the improvement of computer programs that can entrance data and practice it to study for themselves.

**Multi-Layer Perceptron (MLP):** Multi-layer perceptron is also known as a feed forward neural network. It contains of three categories of layers, i.e., 1) the input layer, 2) the output layer, and 3) the hidden layer. Generally, the input layer collects the input signal to be managed.

## NOTE

Normally, the investigational advance creates dependable data but may be exclusive and monotonous; at the same time as the simulation process discloses the fundamental base but may not be dependable. A new option, specifically machine learning, opens up novel prospects for investigate in the field of additive manufacturing. The influence of the machine learning-assisted investigational method has previously talked about in earlier parts. Therefore, we mostly spotlight on machine learning -aided simulations here. A small amount of mechanisms have displayed that the synergy of machine learning and simulation can create the computational speed commands of amount earlier than exact physics-based simulation models in additive manufacturing method. Machine learning can also be leveraged to carry out effective experiment in additive manufacturing investigation. To get better characteristics of an object, it is significant to know the effect of compositional and microstructural characteristics on a specified character by manage of experimentations. But, in perform, it is not practicable to methodically diverge a single microstructure characteristic or alloying substance while maintenance the principles of other apparatus predetermined.

# Chapter 5
# Experimental Investigations and Multi-Objective Optimization of Selective Inhibition Sintering Process Using the Dragonfly Algorithm

**Siva Kumar M.**
*Vel Tech Rangarajan Dr. Sagunthala R&D Institute of Science and Technology, India*

**Rajamani D.**
*Vel Tech Rangarajan Dr. Sagunthala R&D Institute of Science and Technology, India*

**Balsubramanian E.**
*Vel Tech Rangarajan Dr. Sagunthala R&D Institute of Science and Technology, India*

## ABSTRACT

*The chapter focuses on utilizing a hybrid approach of response surface methodology and dragonfly algorithm for investigations and optimization of the selective inhibition sintering (SIS) process to improve the mechanical strengths such as tensile and flexural of fabricated high density polyethylene parts. The layer thickness (LT), heater energy (HE), heater and printer feedrate (HFR & PFR) are considered as the independent variables for the investigation. The SIS experiments are planned and conducted through a response surface methodology-based box-Behnken design approach to fabricate the test specimens. The optimal SIS parameters are obtained through a swarm intelligence metaheuristic technique namely dragonfly algorithm*

DOI: 10.4018/978-1-7998-8516-0.ch005

*(DFA). The optimal parameter settings of LT of 0.102 mm, HE of 28.46 J/mm2, HFR of 3.22 mm/sec, and PFR of 110.49 mm/min are achieved through DFA for improved tensile and flexural strengths of 26.21 MPa and 65.71 MPa, respectively. Further, the prediction ability of DFA was compared with particle swarm optimization algorithm.*

## INTRODUCTION

Additive manufacturing (AM) is a new class of layered manufacturing technique used for fabricating intricate customized components in short span, manifest automation and significantly reduced manufacturing cost (Rajamani & Balasubramanian, 2019a). Therefore, AM can be effectively used in fabricating net shaped components in various fields such as aerospace, automotive, bio-medical and defence industries, etc. (Rajamani & Balasubramanian, 2019b). In the past few decades, several AM techniques such as stereolithography, fused deposition modeling (FDM), selective laser sintering (SLS), laser engineered net shaping (LENS), etc., have been developed to fabricate different parts in various applications.

Each AM technique has its unique features such as processing capabilities, method of fabrication and diversity of raw materials. Among all these techniques, powder-based AM processes are recently used in versatile applications of medium-to-high volume series production (Esakki et al., 2017). Direct metal laser sintering (DMLS), three-dimensional printing (3DP) and Selective laser sintering (SLS) are the most-widely used powder-based AM techniques in fabricating plastic, metal and ceramic parts. The utilization of high-cost heating mechanisms such as laser and electron lead the machine and processing cost. Eliminating laser and electron heating elements in AM processes has incredible impression on reducing the machine cost and speed of the process increased.

In view of this, high speed sintering (HSS) and selective inhibition sintering (SIS) processes are evolved to significantly reduce the processing cost by eliminating the costlier heating elements. HSS involves cost-effective infra-red heater to sinter the powder particles. The heat is transferred to the powder surface through a radiation absorbing material (RAM), which absorbs the as received heat radiation from infra-red heater and transfer it for achieving effective sintering (Majewski et al., 2008). However, the incidence of RAM and more consumption of polymer powder material are foremost challenges to accustom HSS.

To overcome this issue, SIS system was built at University of Southern California, USA (Khoshnevis et al., 2003). In the SIS process, powder particles are sintered with desired part peripheries which is defined through precise delivery of inhibitors. The SIS has the key advantages of eliminating expensive tooling and support structure, processing various indigenously available raw materials such as polymer, metal and ceramics which makes the process cost-effective.

Among the several advantages, SIS has few drawbacks such as compatibility issues with NC tool path generation, compaction of powder particles, effective usage of raw materials and improving the quality characteristics of sintered parts such as strength, surface quality, dimensional stability, etc. (Rajamani, Balasubramanian, Arunkumar et al, 2018). The quality and performance features of the SIS parts can be improved by appropriate selection of process parameters. Several SIS process parameters such as thickness of powder layer, supplied heat energy, part bed temperature, feedrate of heater and inhibitor printer and particle size are having prominent influence on aforementioned functional qualities of sintered parts (Asiabanpour et al., 2007). Hence, it is necessary to investigate the influence of process parameters and identifying optimal values of parameters to improve the quality of sintered parts is very much essential.

Presently various modeling and optimization techniques such as statistical (desirability, GRA and TOPSIS) and metaheuristic approaches (GA, SA, PSO, ABC, etc.,) have been used to enhance the performance of AM process and quality of fabricated parts (Boschetto et al., 2011; Calignano et al., 2012; Gholaminezhad et al., 2016; Mahapatra & Sood, 2011; Raju et al., 2018; Strano et al., 2011). However, statistical optimization techniques solutions are often discrete combinations of profound ranges of process parameters and they may fall on local optima (Javed et al., 2018). Therefore, researchers are keenly looking for appropriate optimization technique to obtain global optimal solutions. Although, several researchers have attempted the hybridization of optimization techniques to arrive at optimal process parameters (Panda et al., 2014; Peng et al., 2014; Sood et al., 2009; Vijayaraghavan et al., 2014).

As can be seen from the existing literatures, there have not been any investigations on implementation of novel metaheuristic optimization technique namely dragonfly algorithm for optimization of any AM process. Moreover, due to the existence of several parameters and non-linearity of SIS process, the dragonfly algorithm (DFA) is adopted to identify the optimal parameters to enhance the mechanical strengths of sintered parts. The experiments were designed and executed by using RSM-BBD approach. The statistical validations of proposed experimental strategy are investigated through analysis of variance. Moreover, the efficiency of utilized DFA is compared with a well-known particle swarm algorithm.

# PROPOSED METHODOLOGY

## Response Surface Methodology

Response surface methodology (RSM) (Ananthakumar et al., 2019; Rajamani, Ananthakumar, Balasubramanian et al, 2018; Tamilarasan & Rajamani, 2017) is employed for modelling and optimizing the SIS governing parameters. The box-behnken design (BBD) is a RSM based approach which is utilized to minimize the experimental runs and quadratic model is established and also the interactions between the SIS governing parameters are studied. The second-order polynomial equation formed from RSM was utilized to manifest the behaviour of the SIS process is given in the Eqn. (1). In this investigation, the surface roughness characteristics of sintered specimens are modelled through accounting the selected process parameters.

$$Y = C_0 + \sum_{i=1}^{n} C_i X_n + \sum_{i=1}^{n} d_i X_i^2 \pm \varepsilon \tag{1}$$

A second-order quadratic model is developed using RSM for correlating process variables and the responses. Additionally, ANOVA is exploited to justify the consequence of developed quadratic models. The SIS process parameters and their levels are decided by conducting several pilot experiments and existing literatures. Selected SIS parameters and their lower and higher levels are described in Table 1.

*Table 1. SIS process parameters and levels*

| Symbols | Parameters | Units | Coded values | | |
|---|---|---|---|---|---|
| | | | -1 (Low) | 0 (medium) | 1 (high) |
| *LT* | Layer thickness | mm | 0.1 | 0.15 | 0.2 |
| *HE* | Heater energy | J/mm$^2$ | 22.16 | 25.32 | 28.48 |
| *HFR* | Heater feedrate | mm/sec | 3 | 3.25 | 3.5 |
| *PFR* | Printer feedrate | mm/min | 100 | 110 | 120 |

## Dragonfly Algorithm

Dragonfly algorithm (DFA) is a nature-inspired optimization approach which was developed based on the migration for food and hunting nature of the dragonflies (Mirjalili, 2015). In recent past, the DFA is prominently used for solving single and multi-objective complex problems. The global optima of dragonfly algorithm are achieved by exploration and exploitation through the static and dynamic swarming behaviour of dragonflies. The mathematical expressions of DFA are described as follows:

The number of dragonflies (*N*) involved in the optimization is represented as,

$$N_i^D = \left(n_1, n_{2,} \ldots n_N\right) \tag{1}$$

where $i = 1, 2, 3, \ldots N$, $N$ is number of the searching flies, and $N_i^D$ is the position of $i^{th}$ dragonfly in the dimensional space of $D$.

The individual behaviours of dragonflies such as separation, alignment, cohesion and alignment can be expressed as follows:

*Separation*: The strategy of separating the individual dragonfly from the other factors is denoted as separation, and it is denoted as,

$$S_i = -\sum_{i=1}^{\overline{N}} Y - Y_i \tag{1}$$

*Alignment*: Alignment is a strategy in which the dragonfly will set its velocity and with respect to the velocity vector of similar dragonflies. It could be explained as,

$$A_i = \frac{\sum_{i=1}^{N} V_i}{N} \tag{1}$$

where $V_i$ indicated the velocity vector of $i^{th}$ adjacent dragonfly.

*Cohesion*: Cohesion implies the characteristics of flies to move in a direction towards the nearby origin of mass. It could be expressed as,

$$C_i = \frac{\sum_{i=1}^{N} n_i}{N} - X \tag{1}$$

*Attraction*: The movement of flies towards the food source is called as attraction and is explained as,

$$F_i = F_{location} - X \tag{1}$$

*Distraction*: The protection of dragonflies from the conflicting enemies is called as distraction, which can be expressed as,

$$E_i = E_{local} + X \tag{1}$$

The step and location vectors can be used for defining the current position of each dragonfly members. They are expressed as follows:

*Step vector*: The direction of dragonfly's movement is called as step vector and it can be expressed by,

$$\Delta X_{t+1} = \left(aA_i + bB_i + cC_i + dD_i\right) + w\Delta X_t \tag{1}$$

where $a, b, c, d, w$ are the weighing factors of each flies.

The dragonfly's location vector can be expressed as,

$$X_{t+1} = X_t + \Delta X_{t+1} \tag{1}$$

Thereafter, the fitness of individual dragonfly is assessed through the position vector and it will be continued until the execution criterion will met.

## EXPERIMENT DETAILS

A custom-built SIS system as shown in Figure 1 has been developed to fabricate the plastic near net shaped end use components. The SIS system consist of following elements: powder feed chamber which supplies the raw powder to the built chamber; part-built chamber to fabricate the desired parts; a recycling chamber to collect the excess powder; an infrared heating element to sinter the powder surface and an inhibition mechanism to define and separate the part boundary. The detailed SIS system development procedure is explained by Balasubramanian et al. (Esakki et al., 2021). According to proposed box behnken experimental design, twenty-nine test samples are fabricated to assess each response characteristics such as tensile and flexural strengths.

The tensile and flexural strengths of the fabricated test samples are assessed through ASTM testing standards and the measured response values are presented in Table 2.

*Figure 1. Custom built SIS system*

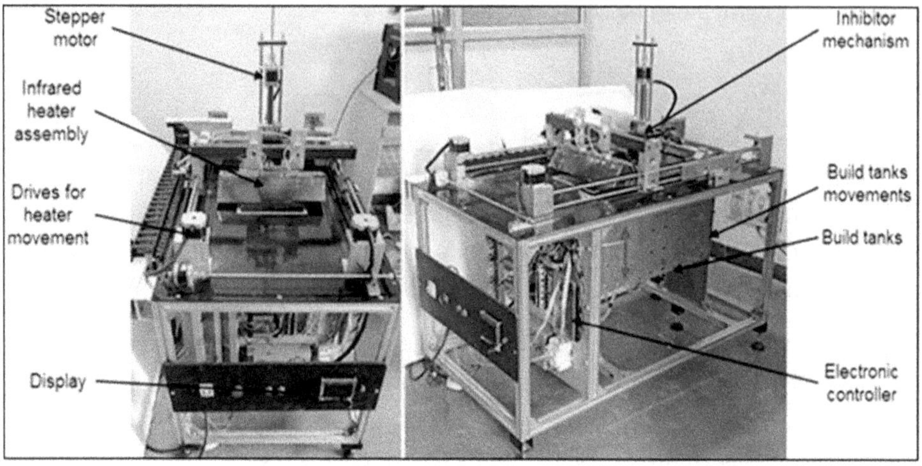


## RESULTS AND DISCUSSIONS

### Statistical Analysis of Regression Models

In order to statistically investigate the proposed experimental strategy and the performed experiments, the multi parameter analysis of variance (ANOVA) is performed along with the residual plots. The ANOVA results for tensile and flexural strengths are described in Table (3-4). The ANOVA indicated that the individual, two-way interaction and the quadratic influences of SIS variables on the selected responses. From the ANOVA results, it is observed that the heater energy is the most influencing variable which mainly dominates the tensile and flexural strengths. The values of coefficient of determination for *TS* and *FS* is found that 95.13% and 86.68%, respectively. Moreover, the model F-value for *TS* is 30.17 and *FS* is 6.51, which indicates the model is adequate.

From the statistical investigations, the insignificant terms of process variables which was not influencing the response values are removed through backward-removal process and the remaining terms are utilized for developing the quadratic mathematical equations. The obtained mathematical terms for *TS* and *FS* are as follows:

$$TS(MPa) = 90.0 - 373.6\,LT - 10.67\,HE + 66.9\,HFR - 0.138\,PFR - 0.0380\,HE*HE$$
$$-32.64\,HFR*HFR - 0.00928\,PFR*PFR + 110.6\,LT*HFR + 3.089\,HE*HFR$$
$$+0.02603\,HE*PFR + 0.447\,HFR*PFR \tag{1}$$

*Table 2. Experimental layout using RSM-BBD and output response values*

| Standard run no. | Factors | | | | | Responses | |
|---|---|---|---|---|---|---|---|
| | *LT* (mm) | *HE* (J/ mm²) | *HFR* (mm/ sec) | *PFR* (mm/ min) | | Tensile Strength (MPa) | Flexural Strength (MPa) |
| 1 | 0.1 | 22.16 | 3.25 | 110 | | 23.8 | 48.01 |
| 2 | 0.2 | 22.16 | 3.25 | 110 | | 23.77 | 52.57 |
| 3 | 0.1 | 28.48 | 3.25 | 110 | | 26.44 | 64.88 |
| 4 | 0.2 | 28.48 | 3.25 | 110 | | 24.97 | 48.94 |
| 5 | 0.15 | 25.32 | 3 | 100 | | 24.97 | 53 |
| 6 | 0.15 | 25.32 | 3.5 | 100 | | 21.62 | 45.4 |
| 7 | 0.15 | 25.32 | 3 | 120 | | 20.41 | 43.11 |
| 8 | 0.15 | 25.32 | 3.5 | 120 | | 21.53 | 47.2 |
| 9 | 0.1 | 25.32 | 3.25 | 100 | | 24.85 | 52.11 |
| 10 | 0.2 | 25.32 | 3.25 | 100 | | 24.39 | 46.35 |
| 11 | 0.1 | 25.32 | 3.25 | 120 | | 24.94 | 47.18 |
| 12 | 0.2 | 25.32 | 3.25 | 120 | | 22.9 | 35.77 |
| 13 | 0.15 | 22.16 | 3 | 110 | | 24.47 | 47.71 |
| 14 | 0.15 | 28.48 | 3 | 110 | | 21.29 | 56.64 |
| 15 | 0.15 | 22.16 | 3.5 | 110 | | 19.34 | 39.24 |
| 16 | 0.15 | 28.48 | 3.5 | 110 | | 25.92 | 53.57 |
| 17 | 0.1 | 25.32 | 3 | 110 | | 25.87 | 57.6 |
| 18 | 0.2 | 25.32 | 3 | 110 | | 20.86 | 44.18 |
| 19 | 0.1 | 25.32 | 3.5 | 110 | | 22.53 | 36.34 |
| 20 | 0.2 | 25.32 | 3.5 | 110 | | 23.05 | 40.46 |
| 21 | 0.15 | 22.16 | 3.25 | 100 | | 23.97 | 48.9 |
| 22 | 0.15 | 28.48 | 3.25 | 100 | | 24.64 | 53.13 |
| 23 | 0.15 | 22.16 | 3.25 | 120 | | 21.29 | 45.04 |
| 24 | 0.15 | 28.48 | 3.25 | 120 | | 25.25 | 54.19 |
| 25 | 0.15 | 25.32 | 3.25 | 110 | | 24.7 | 52.1 |
| 26 | 0.15 | 25.32 | 3.25 | 110 | | 25.29 | 55.22 |
| 27 | 0.15 | 25.32 | 3.25 | 110 | | 25.12 | 54.55 |
| 28 | 0.15 | 25.32 | 3.25 | 110 | | 24.9 | 52.22 |
| 29 | 0.15 | 25.32 | 3.25 | 110 | | 25.55 | 55.63 |

*Table 3. ANOVA for tensile strength*

| Source | DF | Adj. SS | Adj. MS | F-Value | P-Value |
|---|---|---|---|---|---|
| *Model* | 11 | 94.1773 | 8.5616 | 30.17 | 0.000 |
| *Linear* | 4 | 24.4971 | 6.1243 | 21.58 | 0.000 |
| LT | 1 | 6.0067 | 6.0067 | 21.17 | 0.000 |
| HE | 1 | 11.7414 | 11.7414 | 41.37 | 0.000 |
| HFR | 1 | 1.2545 | 1.2545 | 4.42 | 0.051 |
| PFR | 1 | 5.4945 | 5.4945 | 19.36 | 0.000 |
| *Square* | 3 | 30.5193 | 10.1731 | 35.85 | 0.000 |
| HE*HE | 1 | 0.9662 | 0.9662 | 3.40 | 0.083 |
| HFR*HFR | 1 | 28.0027 | 28.0027 | 98.67 | 0.000 |
| PFR*PFR | 1 | 5.7901 | 5.7901 | 20.40 | 0.000 |
| *2-Way Interaction* | 4 | 39.1609 | 9.7902 | 34.50 | 0.000 |
| LT*HFR | 1 | 7.6452 | 7.6452 | 26.94 | 0.000 |
| HE*HFR | 1 | 23.8144 | 23.8144 | 83.91 | 0.000 |
| HE*PFR | 1 | 2.7060 | 2.7060 | 9.53 | 0.007 |
| HFR*PFR | 1 | 4.9952 | 4.9952 | 17.60 | 0.001 |
| *Error* | 17 | 4.8246 | 0.2838 | | |
| Lack-of-Fit | 13 | 4.3863 | 0.3374 | 3.08 | 0.144 |
| Pure Error | 4 | 0.4383 | 0.1096 | | |
| *Total* | 28 | 99.0019 | | | |
| $R^2$ | 95.13 | *Adj.* $R^2$ | 91.97 | | |

$$FS(MPa) = -478 + 357LT - 9.9HE + 263HFR + 4.10PFR - 1428LT*LT$$
$$+0.124HE*HE - 77.0HFR*HFR - 0.0395PFR*PFR - 32.4LT*HE + 351LT*HFR$$
$$-2.83LT*PFR + 1.71HE*HFR + 0.0389HE*PFR + 1.169HFR*PFR \quad (2)$$

In addition to the ANOVA, the residual plots are proposed and investigated for the statistical validation of performed experimental studies. The residual plots for TS and FS are presented in Figure 2. The normal probability plot, versus fits, histogram and versus order are considered as the performance indices for statistical validation. From the Figure, it is clearly observed that the actual data are distributed normally with minimal deviations from the experimental data, and hence the obtained mathematical models can be effectively used for further investigations.

*Table 4. ANOVA for flexural strength*

| Source | DF | Adj. SS | Adj. MS | F-Value | P-Value |
|---|---|---|---|---|---|
| *Model* | 14 | 1060.93 | 75.781 | 6.51 | 0.001 |
| *Linear* | 4 | 518.33 | 129.583 | 11.13 | 0.000 |
| LT | 1 | 119.39 | 119.385 | 10.26 | 0.006 |
| HE | 1 | 207.33 | 207.335 | 17.81 | 0.001 |
| HFR | 1 | 133.53 | 133.533 | 11.47 | 0.004 |
| PFR | 1 | 58.08 | 58.080 | 4.99 | 0.042 |
| *Square* | 4 | 305.14 | 76.285 | 6.55 | 0.003 |
| LT*LT | 1 | 82.67 | 82.666 | 7.10 | 0.018 |
| HE*HE | 1 | 9.87 | 9.875 | 0.85 | 0.373 |
| HFR*HFR | 1 | 150.22 | 150.223 | 12.90 | 0.003 |
| PFR*PFR | 1 | 101.01 | 101.009 | 8.68 | 0.011 |
| *2-Way Interaction* | 6 | 237.46 | 39.577 | 3.40 | 0.028 |
| LT*HE | 1 | 105.06 | 105.063 | 9.03 | 0.009 |
| LT*HFR | 1 | 76.91 | 76.913 | 6.61 | 0.022 |
| LT*PFR | 1 | 7.98 | 7.981 | 0.69 | 0.422 |
| HE*HFR | 1 | 7.29 | 7.290 | 0.63 | 0.442 |
| HE*PFR | 1 | 6.05 | 6.052 | 0.52 | 0.483 |
| HFR*PFR | 1 | 34.16 | 34.164 | 2.93 | 0.109 |
| *Error* | 14 | 162.97 | 11.641 | | |
| Lack-of-Fit | 10 | 151.76 | 15.176 | 5.41 | 0.059 |
| Pure Error | 4 | 11.21 | 2.803 | | |
| Total | 28 | 1223.91 | | | |
| $R^2$ | 86.68 | *Adj.* $R^2$ | 73.37 | | |

*Figure 2. Residual plots for statistical validation of tensile and flexural strengths*

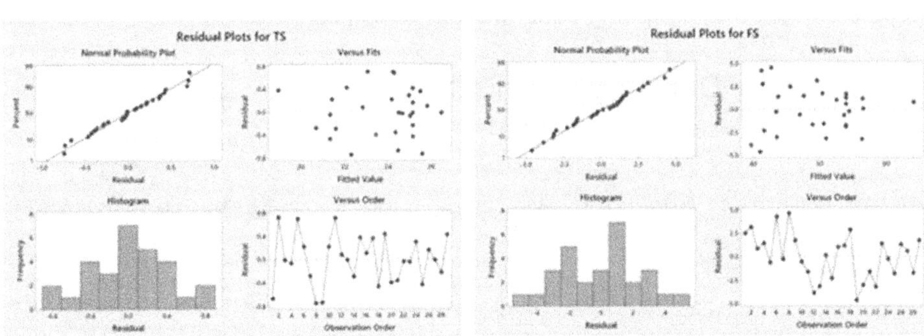

## Multi-Response Optimization through Dragonfly Algorithm (DFA)

The objective of present study is to maximize the tensile and flexural strength of SIS fabricated HDPE parts with respect to the selected bounds of processing parameters. For this purpose, a population-based metaheuristic evolutionary optimization approach called Dragonfly optimization algorithm has been used. The DFA is coded and executed in Matlab 2019b® environment. According to the Pseudo code of DFA, the algorithm is executing. The second order multiple linear regression models obtained through experimental data were utilized as objective functions for DFA to obtain optimal SIS parameters. The optimization problem of this present investigation is summarized as follows:

$$Maximize = Strength\ characteristics\left(Tensile\ \&\ Flexural\right)$$

$$Subjected\ to$$

$$0.1 \leq LT \leq 0.2$$

$$22.16 \leq HE \leq 28.48$$

$$3 \leq HFR \leq 3.5$$

$$100 \leq PFR \leq 120$$

The DFA was initialized with the parameters as given in Table 5 to obtain feasible optimal solutions to improve the tensile and flexural strengths of SIS processed parts. In general, the combination of a process parameter set for an objective function is not appropriate for any more objective functions. Such complex optimization problems are typically solved through two approaches. The first one is to convert multi-objectives into a single objective function by assigning weights or utility function for each objective. The second one is to obtain non-dominated Pareto optimal settings of decision variables (Ciurana et al., 2009). In this study, Deng's method, a statistical approach, has been adopted to convert the multi-objective functions into a single-objective by assigning equal weights to respective responses.

### Pseudocode of Dragonfly Optimization Algorithm

```
Define number of dragonflies (nd), number of iteration (nitr),
and archive size
As initial population, initialize position of dragonflies Pij
Assign step vector (Vij) values as Pij
```

*Table 5. Dragonfly algorithm parameters for optimization*

| Parameters | Value/Equation |
|---|---|
| No. of dragonflies (*nd*) | 100 |
| No. of Iterations (*nitr*) | 100 |
| Inertia weight (*w*) ($w_{max}$=0.9 and $w_{min}$=0.2) | $w = w_{max} - \left(w_{max} - w_{min}\right)\frac{itr}{nitr}$ |
| Separation weight | $sw = 0.1 - \frac{0.1 * itr}{nitr}$ |
| Alignment weight | $aw = 0.1 - \frac{0.1 * itr}{nitr}$ |
| Cohesion weight | $cw = 0.1 - \frac{0.1 * itr}{nitr}$ |
| Food factor | |
| Enemy factor | $ef = 0.1 - \frac{0.1 * itr}{nitr}$ |
| Achieve size | 100 |

```
Do While i<=nitr
        Calculate the value of inertia, separation, alignment
and cohesion weights,
            food and enemy factor values as per the equation
mentioned in the above Table 5.
Compute the objective values of each dragonflies (Fi)
        Determine the non-dominated objective values
        Update the no. of non-dominated solutions in archive
        Assume best solution as food source and worst solution
as enemy
        Update the values of Vij and Pij
        Check Pij values lies between the lower and upper
limits of process parameters
End
```

Convert the non-dominated objective values into single objective value using Deng's method

*Figure 3. Pareto optimal front obtained through dragonfly algorithm and particle swarm optimization*

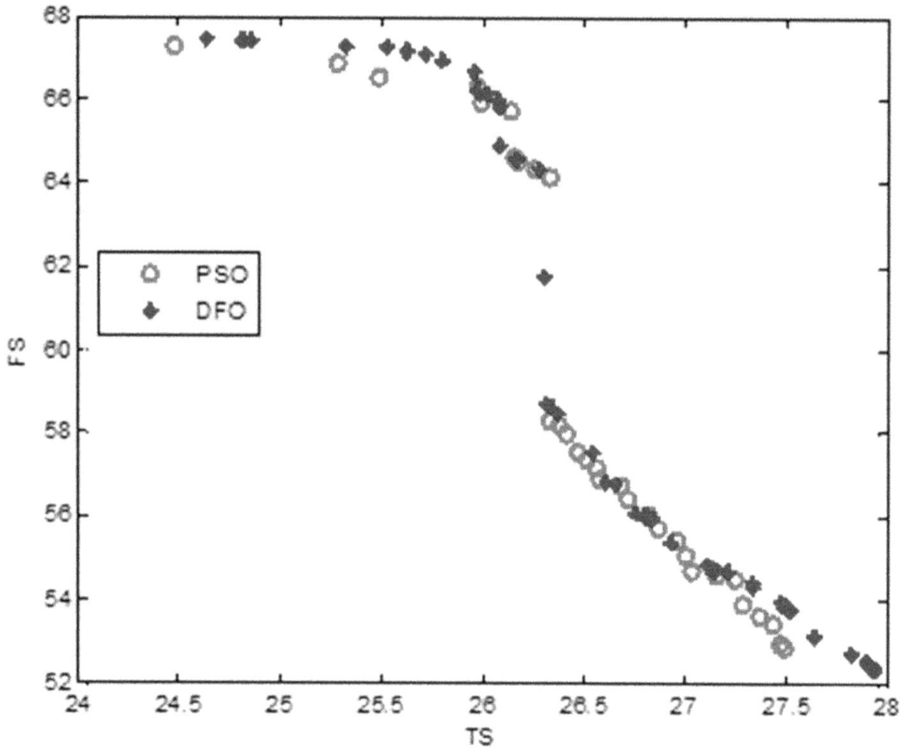


Display the best objective values with its optimum process parameter's values.

The obtained pareto optimal front for simultaneous optimization of tensile and flexural strength characteristics through DFA and PSO is indicated in Figure 3. The optimization was performed several times and the optimal set of parameters for each run has been taken for further investigations. Through the several optimization trials, a range of optimal SIS parameters were obtained. Further, the multi-objective functions were converted to single objective through Deng's approach for obtaining a single set of global optimal solutions among the obtained pareto optimal solutions. Figure 4(a-b) shows the convergence plots of DFA and PSO for the mechanical strength characteristics. From the plot, it is observed that the algorithm is converged at 43$^{rd}$ iteration number for DFA, whereas PSO is converged at 54$^{th}$ iteration number for optimizing tensile strength. Similarly, for the flexural strength, the DFA provides optimal solution at 12$^{th}$ iteration, whereas the PSO converged at 37$^{th}$ iteration. Hence, it is proved that the DFA can provide near optimal solutions with minimal iterations to compare the well-known PSO technique. The obtained optimal SIS parameters and their corresponding tensile and flexural strengths for both DFA and PSO algorithms are shown in Table 6.

*Figure 4. Convergence plots for tensile and flexural strength characteristics*

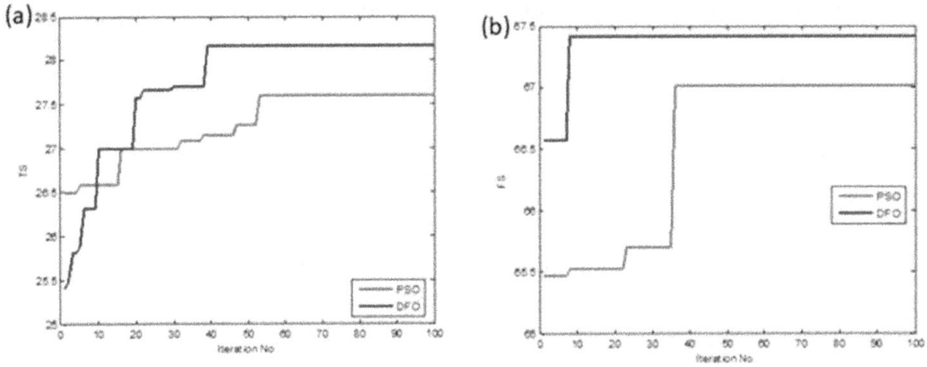


*Table 6. Optimal SIS parameter obtained using DFO and PSO*

| Algorithm | LT | HE | HFR | PFR | TS | FS |
|---|---|---|---|---|---|---|
| PSO | 0.1019 | 28.41 | 3.25 | 109.92 | 26.32 | 64.85 |
| DFO | 0.1019 | 28.45 | 3.22 | 110.49 | 26.21 | 65.71 |

## CONCLUSION

The present chapter indented to dealt the optimization of novel AM process namely selective inhibition sintering parameters through metaheuristic algorithms. The results of presented work are summarized as follows:

- The SIS experimental specimens are successfully fabricated using RSM-BBD design strategy and the statistical investigations are performed through ANOVA.
- The statistical results indicated that the heater energy is most significant parameter on the tensile and flexural strength characteristics.
- The optimal SIS variables for improved tensile and flexural strengths are obtained through a metaheuristic dragonfly algorithm. The obtained optimal variables are: LT of 0.1019 mm, HE of 28.45 J/mm$^2$, HFR of 3.22 mm/sec, and PFR of 110.49 mm/min, respectively.

- The performance of dragonfly algorithm is assessed through comparing its results with particle swarm optimization algorithm and found that the dragonfly algorithm is performing better in prediction of optimal SIS parameters in terms of convergence rate.
- Therefore, the proposed dragonfly algorithm can be effectively cast-off in prediction of optimal parameters for similar additive manufacturing processes.

## REFERENCES

Ananthakumar, K., Rajamani, D., Balasubramanian, E., & Paulo Davim, J. (2019). Measurement and optimization of multi-response characteristics in plasma arc cutting of Monel 400™ using RSM and TOPSIS. *Measurement*, *135*, 725–737. doi:10.1016/j.measurement.2018.12.010

Asiabanpour, B., Cano, R., Subbareddy, C., Wasik, F., VanWagner, L., & McCormick, T. (2007). A new heater design by radiation modeling and a new polymer waste-saving mechanism design for the SIS process. *Rapid Prototyping Journal*, *13*(3), 136–147. doi:10.1108/13552540710750889

Boschetto, A., Giordano, V., & Veniali, F. (2011). Modelling micro geometrical profiles in fused deposition process. *International Journal of Advanced Manufacturing Technology*, *61*(9-12), 945–956. doi:10.100700170-011-3744-1

Calignano, F., Manfredi, D., Ambrosio, E. P., Iuliano, L., & Fino, P. (2012). Influence of process parameters on surface roughness of aluminum parts produced by DMLS. *International Journal of Advanced Manufacturing Technology*, *67*(9-12), 2743–2751. doi:10.100700170-012-4688-9

Ciurana, J., Arias, G., & Ozel, T. (2009). Neural Network Modeling and Particle Swarm Optimization (PSO) of Process Parameters in Pulsed Laser Micromachining of Hardened AISI H13 Steel. *Materials and Manufacturing Processes*, *24*(3), 358–368. doi:10.1080/10426910802679568

Esakki, B., Ali, T. K., Rajamani, D., & Sachin, S. (2021). Parametric Optimization on Impact Strength of Selective Inhibition Sintering Fabricated PA-12 Parts Based on Evolutionary Optimization Algorithms. *Journal of Materials Engineering and Performance*, *30*(7), 5356–5367. Advance online publication. doi:10.100711665-021-05794-1

Esakki, B., Rajamani, D., & Arunkumar, P. (2017). Modeling and prediction of optimal process parameters in wear behaviour of selective inhibition sintered high density polyethylene parts. *Progress in Additive Manufacturing*, *3*(3), 109–121. doi:10.100740964-017-0033-z

Gholaminezhad, I., Assimi, H., Jamali, A., & Vajari, D. A. (2016). Uncertainty quantification and robust modeling of selective laser melting process using stochastic multi-objective approach. *International Journal of Advanced Manufacturing Technology*, *86*(5-8), 1425–1441. doi:10.100700170-015-8238-0

Javed, S., Mahmoudi, A., & Khan, A. M. (2018). Investigation of drilling parameters on hybrid polymer composites using grey relational analysis, regression, fuzzy logic, and ANN models: A critical note. *Journal of the Brazilian Society of Mechanical Sciences and Engineering*, *40*(12), 560. doi:10.100740430-018-1470-4

Khoshnevis, B., Asiabanpour, B., Mojdeh, M., & Palmer, K. (2003). SIS - a new SFF method based on powder sintering. *Rapid Prototyping Journal*, *9*(1), 30–36. doi:10.1108/13552540310455638

Mahapatra, S. S., & Sood, A. K. (2011). Bayesian regularization-based Levenberg–Marquardt neural model combined with BFOA for improving surface finish of FDM processed part. *International Journal of Advanced Manufacturing Technology*, *60*(9-12), 1223–1235. doi:10.100700170-011-3675-x

Majewski, C. E., Oduye, D., Thomas, H. R., & Hopkinson, N. (2008). Effect of infra-red power level on the sintering behaviour in the high-speed sintering process. *Rapid Prototyping Journal*, *14*(3), 155–160. doi:10.1108/13552540810878012

Mirjalili, S. (2015). Dragonfly algorithm: A new meta-heuristic optimization technique for solving single-objective, discrete, and multi-objective problems. *Neural Computing & Applications*, *27*(4), 1053–1073. doi:10.100700521-015-1920-1

Panda, B. N., Bahubalendruni, M. V. A. R., & Biswal, B. B. (2014). A general regression neural network approach for the evaluation of compressive strength of FDM prototypes. *Neural Computing & Applications*, *26*(5), 1129–1136. doi:10.100700521-014-1788-5

Peng, A., Xiao, X., & Yue, R. (2014). Process parameter optimization for fused deposition modeling using response surface methodology combined with fuzzy inference system. *International Journal of Advanced Manufacturing Technology*, *73*(1-4), 87–100. doi:10.100700170-014-5796-5

Rajamani, D., Ananthakumar, K., Balasubramanian, E., & Paulo Davim, J. (2018). Experimental investigation and optimization of PAC parameters on Monel 400™ superalloy. *Materials and Manufacturing Processes*, *33*(16), 1864–1873. doi:10.1080/10426914.2018.1532085

Rajamani, D., & Balasubramanian, E. (2019a). Investigations of sintering parameters on viscoelastic behaviour of selective heat sintered HDPE parts. *Journal of Applied Science and Engineering*, *22*(3), 391–402.

Rajamani, D., & Balasubramanian, E. (2019b). Effects of heat energy on morphology and properties of selective inhibition sintered high density polyethylene. *Journal of Mechanical Engineering Science*, *13*(1), 4403–4414. doi:10.15282/jmes.13.1.2019.05.0375

Rajamani, D., Balasubramanian, E., Arunkumar, P., Silambarasan, M., & Bhuvaneshwaran, G. (2018). Experimental Investigations and Parametric Optimization of Process Parameters on Shrinkage Characteristics of Selective Inhibition Sintered High Density Polyethylene Parts. *Experimental Techniques*, *42*(6), 631–644. doi:10.100740799-018-0286-6

Raju, M., Gupta, M. K., Bhanot, N., & Sharma, V. S. (2018). A hybrid PSO–BFO evolutionary algorithm for optimization of fused deposition modelling process parameters. *Journal of Intelligent Manufacturing*, *30*(7), 2743–2758. doi:10.100710845-018-1420-0

Sood, A. K., Ohdar, R. K., & Mahapatra, S. S. (2009). Parametric appraisal of fused deposition modelling process using the grey Taguchi method. *Proceedings of the Institution of Mechanical Engineers. Part B, Journal of Engineering Manufacture*, *224*(1), 135–145. doi:10.1243/09544054JEM1565

Strano, G., Hao, L., Everson, R. M., & Evans, K. E. (2011). Multi-objective optimization of selective laser sintering processes for surface quality and energy saving. *Proceedings of the Institution of Mechanical Engineers. Part B, Journal of Engineering Manufacture*, *225*(9), 1673–1682. doi:10.1177/0954405411402925

Tamilarasan, A., & Rajamani, D. (2017). Multi-response optimization of Nd: YAG laser cutting parameters of Ti-6Al-4V superalloy sheet. *Journal of Mechanical Science and Technology*, *31*(2), 813–821. doi:10.100712206-017-0133-1

Vijayaraghavan, V., Garg, A., Lam, J. S. L., Panda, B., & Mahapatra, S. S. (2014). Process characterisation of 3D-printed FDM components using improved evolutionary computational approach. *International Journal of Advanced Manufacturing Technology*, *78*(5-8), 781–793. doi:10.100700170-014-6679-5

## KEY TERMS AND DEFINITIONS

**3DP:** Three-dimesnional printing.
**ABC:** Artificial bee-colony algorithm.
**AM:** Additive manufacturing.
**ANOVA:** Analysis of variance.
**BBD:** Box Behnken design.
**DFA:** Dragonfly algorithm.
**DMLS:** Direct metal laser sintering.
**FDM:** Fused deposition modelling.
**GA:** Genetic algorithm.
**GRA:** Grey relational analysis.
**HSS:** High speed sintering.
**LENS:** Laser engineered net shaping.
**PSO:** Particle swarm optimization.
**RAM:** Radiation absorbing materials.
**RSM:** Response surface methodology.
**SIS:** Selective inhibition sintering.
**SLS:** Selective laser sintering.
**TOPSIS:** Technique for order of preference by similarity to ideal solution.

# Chapter 6
# Optimized Robotic WAAM

**Aya Abd Alla Ramadan**
*Faculty of Engineering, Helwan University, Egypt*

**Sherif Elatriby**
*Faculty of Engineering, Helwan University, Egypt*

**Abd El Ghany**
*Faculty of Engineering, Helwan University, Egypt*

**Azza Fathalla Barakat**
*Faculty of Engineering, Helwan University, Egypt*

## ABSTRACT

*This chapter summarizes a PhD thesis introducing a methodology for optimizing robotic MIG (metal inert gas) to perform WAAM (wire and arc additive manufacturing) without using machines equipped with CMT (cold metal transfer) technology. It tries to find the optimal MIG parameters to make WAAM using a welding robot feasible production technique capable of making functional products with proper mechanical properties. Some experiments were performed first to collect data. Then NN (neural network) models were created to simulate the MIG process. Then different optimization techniques were used to find the optimal parameters to be used for deposition. These results were practically tested, and the best one was selected to be used in the third stage. In the third stage, a block of metal was deposited. Then samples were cut from deposited blocks in two directions and tested for tension stress. These samples were successful. They showed behavior close to base alloy.*

DOI: 10.4018/978-1-7998-8516-0.ch006

## INTRODUCTION

Metal 3D printing faces a lot of problems although it is very important for actual prototyping. In case laser printers were used there will be a lot powder shedding, it is a very time-consuming process, powders materials are very limited and dimensions applicable are very limited. Also, in case of using MIG (metal inert gas) WAAM (wire and arc additive manufacturing) it requires applying CMT (cold metal transfer to get difficult geometries and avoid heat effects on the product.

Here we will discuss optimizing robotic MIG WAAM in order to get a metal 3D printed product avoiding defects resulting from arc heat. we tried to find the optimal MIG parameters to make WAAM using welding robot feasible production technique capable of making functional products with proper mechanical properties.

## BACKGROUND

AM (additive manufacturing) idea has evolved about 29 years ago different materials and different power sources have been used for it (Salonitis, 2016). WAAM (wire and arc additive manufacturing) has evolved few years ago. The most popular wire arc method used for WAAM is MIG welding / GMAW (gas metal arc welding). One of the main challenges facing WAAM is obtaining a product with sufficient mechanical properties to be functional, require less machining after deposition and need no heat treatment to be able to merge the technique with regular production methods.

To achieve this target it was necessary to model the MIG WAAM process and find the optimal MIG parameters. Optimization target was to be smallest homogeneous width, homogenous height and suitable hardness values to avoid cracks at all zones. For such a complicated MIMO (multi input multi output) highly nonlinear process like that NN (neural networks) is the best choice to model and understand relation between inputs and outputs. Parameters to be measured and modelled were selected depending of previous work of a lot of researchers, information provided by ASM handbook and observation of experimental specimens. Parameters and their effect on the resulting line are summarized in table.1.

A.Sumesh et al (Sumesh et al., 2017) sensed voltage and current of defected and non-defected welds to make a statistical model enabling defects avoidance in industry. Jorge Giron Cruz et al (Cruz et al., 2015) used NN to model GMAW process taking welding speed; wire feed velocity and arc voltage as inputs and width obtained through image processing system as output. Fuzzy controller is then used to control the weld quality and weld width. It was found that welding speed is the most preferred and affecting parameter to be controlled without affecting metal transfer behavior.

Donghong Ding et al (Ding et al., 2016) developed automated manufacturing system of WAAM (wire and arc additive manufacturing) to automatically design deposition path. First an NN model was developed then used MAT (medial axis transformation) algorithm. Results of this research were combined with a previous research which developed a model of multi bead overlapping. According to this research the best overlap between deposited lines to get nearly flat surface of the deposited layer is 0.738*w, where w is the single width and the distance is measured between centerlines of the sided lines. Also this research confirmed that the contour path starting from outside to inside is the best deposition pattern.

L. Nele et al (Nele et al., 2013) used neuro-fuzzy to model and adaptively control arc welding parameters to get sound weld bead. Wire feed speed, welding gap and torch speed were considered as inputs and welding current was considered as output this model was used for controlling process parameters. The same modeled to predict final weld joint characteristics like dilution ratio, hardness of weld bead, hardness of fused zone and bead width. Other models developed using multiple regressions were developed to be compared with neuro-fuzzy models. Each output variable had its own models developed by neuro-fuzzy and multiple regressions.

Jun Xiong et al (Xiong et al., 2013) made NN model between welding process parameters and bead geometry then a reverse model was developed to construct closed loop iteration system. GMAW is widely used for rapid prototyping because of its high productivity, low cost, high density and excellent bonding strength of components. NN is the best tool for modeling such complex nonlinear process like GMAW. Welding speed, wire feed rate, arc voltage and nozzle to plate distance were chosen as inputs while bead width and bead height were selected as output monitored by vision system. Wire feed was decreased and torch speed was increased to lessen heat input. In case there is difference between forward and reverse models or values of parameters are out of range adjustment is performed. If predicted height and width are both larger than desired welding speed is to be increased or wire feed is to be decreased. If width is larger than desired while height is less than desired voltage is to be slightly decreased.

According to Young Cao et al (Cao et al., 2011) robotic MAG (metal active gas) is a promising technology for rapid prototyping. Single bead geometry and overlap distance is very important for dimension accuracy. Canny edge detection of beads was performed then results were fitted by Gaussian function, Logistic function, Parabola function and sine function. Mathematical model of the bead geometry was developed to analyze it. The optimal overlap was experimentally calculated to be 63.66%. Gaussian function showed lowest accuracy compared to the other 3 functions fitting the measured data while sine was the highest.

Nature inspired optimization techniques proved to be most suitable for optimizing the more and more complicating industrial systems. Roben Lostado-Lorza et al (Lostado-lorza & Fernández-martínez, 2014) say FEM (finite element method) is very computationally costing so GA (genetic algorithm) combined with MT (model trees) were used to improve its performance while dealing with such complex system like GMAW to find the optimal parameters to give optimal bead.

Rakesh Malviya and Dilip Kumar Pratihar (Malviya & Pratihar, 2011) found that RBFNN (radial basis function neural network) is the best tool to model GMAW process with its high nonlinearity. GA (genetic algorithm) is good for global search while PSO (particle swarm optimization) is capable of both global and local search. Their research aimed to optimizing GMAW using NN and modified PSO. Here PSO was used for global optimization and back propagation algorithm was used for local finer optimization. 4 different algorithms were used to test the modified optimizer.

N.Raghavendra et al (Raghavendra et al., 2009) used two techniques to predict weld ultimate tensile strength for pulsed MIG welding. First was back-propagation NN the second was hybrid neuro ant colony. Data was collected through large number of experiments to make both models. Results showed that hybrid neuro ant colony optimized model performs faster and predicted better weld joints than the backpropagation NN model. Inputs for these models were voltage, pulse voltage, pulse frequency, pulse duty factor, wire feed rate and linear speed.

For optimization nature inspired optimization techniques used were GA (genetic algorithm), ACO (ant colony optimization), HS (harmony search) and MACO (modified ant colony optimization) were used.

## Experimental Work

Experiments were performed using KUKA robot equipped with fronious automatic welding machine to ensure accurate distances and offsets. First stage of experiments was single lines deposited of strips of (steel 37) 100×30×4 mm to satisfy the requirements of local factories in Egypt. Width, height, penetration and hardness on three levels were measured in this stage. This data was used to construct the NN model. After optimizing the model performance it was practically verified and the best result was used for accumulating blocks of metal. Then tension specimens were machined from them to test research results.

## Model

To deal with a process with such high nonlinearity and has multi input and multi output it was essential to make a model of it. NN backpropagation proved it is the best method for modelling such process. Single line samples were laid to collect data for modelling the process as a first stage of the research. Parameters used for MIG process and their ranges were selected according to AWS, Miller Company's android application and Fronious IOS application, they are summarized in table 2.

Feed rate was supposed to be one of the controlled parameters, but it was not practically possible as the welding machine changes it according to material, work piece thickness, volt and current. Experiments were designed using Taguchi so there were 3 parameters sets to be performed each one repeated 3 times. Experiments variables were arc current (amber), voltage (volt) and travel speed applied by robot (m/sec). parameters measured from deposited lines were width (mean, standard deviation and variance) because some lines were not homogenous in width as shown in fig.1, height and penetration after cutting the deposited line

*Table 1. Summary of MIG parameters and their effects:*

| | Voltage | Current | Linear velocity | Gas pressure | Gas Mixture | Preheating |
|---|---|---|---|---|---|---|
| **Droplet size** | $I_{Arc} \alpha\alpha \frac{1}{V_{Arc}}$<br>*Droplet* size $\alpha$ $V_{Arc}$ | Droplet size $\alpha$ $\frac{1}{I_{Arc}}$ | | | Add $O_2$ decreases size | |
| **Detachment** | | Frequency of detachment $\alpha$ $I_{Arc}$ | | | | |
| **Spraying** | Spraying $\alpha$ 1/V | | | | Add $O_2$ increase No. of drops per time | |
| **Width** | Width $\alpha$ $V_{Arc}$ | | | | | |
| **Penetration (take care of momentum stream)** | | P $\alpha$ $I_{Arc}$ | | Direct proportional | Argon + $CO_2$ make higher penetration. Argon make deeper narrower than helium. | |
| **HAZ** | | | | | | Inverse proportional (evidence) |
| **Hardness** | | | Direct proportionality | | | |
| **Concaveness** | Concaveness $\alpha$ V | | | | | |
| **Notes** | 1- Increasing penetration increases possibility of burn through and solidification cracks.<br>2- Welding with too low Arc voltage produces a low quality bead.<br>Adding $CO_2$ to Argon increases possibility of base material distortion & lack of gap bridging. | | | | | |

*Table 2. MIG welding parameters*

| | |
|---|---|
| Wire type | ER70S-6 |
| Current | 180: 190 Amp |
| Voltage | 24: 25 Volts |
| CTWD | 12.7mm |
| Travel speed | 10: 125 cm/min |
| Feed rate | 360: 380 IPM |
| Gas Mixture | Argon 75% and CO2 25% |

*Figure 1. Specimen of the first stage experiments with non-homogenous width.*

*Figure 2. Etched horizontal cut of specimen.*

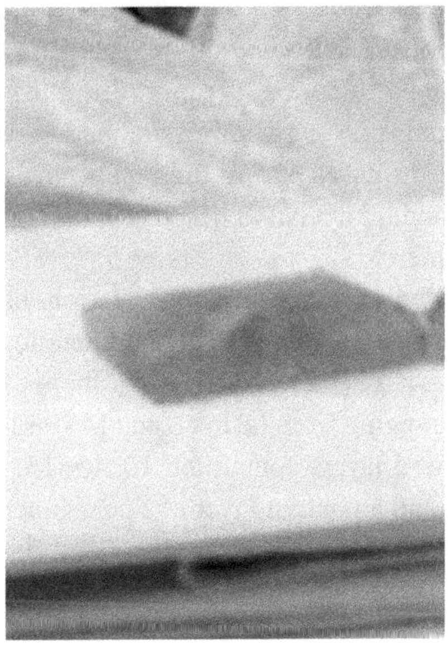

*Figure 3. Hardness specimen.*

horizontally 10 mm away from the start and etching it as shown in fig.2. the rest of the specimen was cut longitudinally, one part of them was etched and hardness measured. Fig.3 shows hardness specimen.

First all MIG parameters were used as inputs and all measured parameters were used as outputs for NN model of the process. Resulting networks were tested by the reserved specimens' measurements, they showed tolerance of ±2% maximum with regression ranging between 76 to 90% and a MSE (mean square error) between 0.23 and 1.13. These results were not satisfying so the model was divided into networks each one resembling the relation between all inputs and one output (Barakat et al., 2018). Fig.4 shows the resulting models.

For width, networks showed regression more than 99 and MSE ranging between 0.16 and 0.32. And all these networks passed the ±2% test. Best network consisted of 2 hidden layers containing 5 and 4 neurons respectively with "tansig" functions and an output layer of linear function. 'Trainlm' training function gave the best result for width. Fig.5 and fig.6 show regression and mean square errors for the best width network, these figures ensure there is no overfitting. This networks' outputs are mean and standard deviation because some lines width showed strong variation for the same line.

For penetration, networks modeling the relation between all inputs and penetration did not pass the ±2% test at all so the tolerance was increased to ±5% but it was not sufficient so it was increased to ±10% then some networks passed the test showing regression ranging from 68 to 86 and MSE ranged from 0.06 to 0.23. best network consisted of 5 layers containing 10, 15, 15, 25 and 30 respectively with "tansig" functions and an output layer of linear function. 'Traingd' training function gave the best result for penetration. Fig.7 and fig.8 show regression and

*Figure 4. Models of the process made in the first stage*

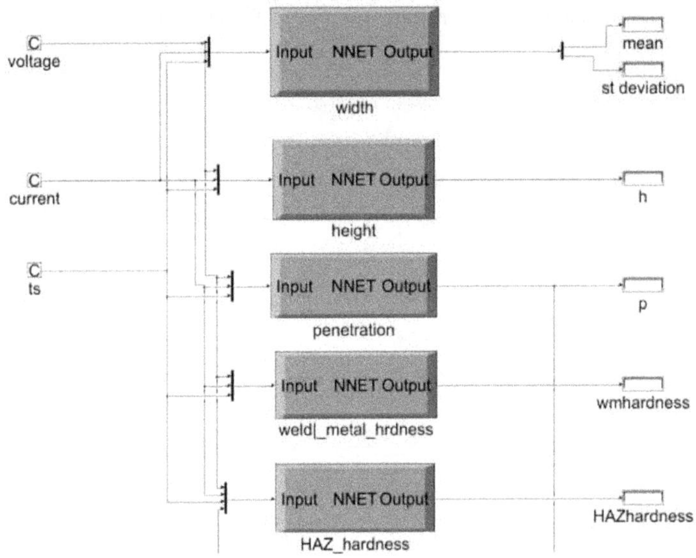


*Figure 5. Best width network regression.*

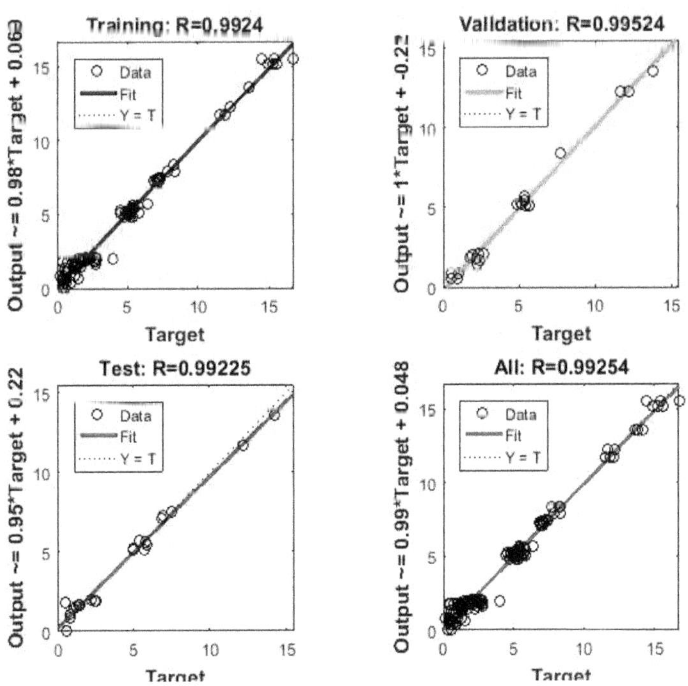

*Figure 6. Best width network MSE.*

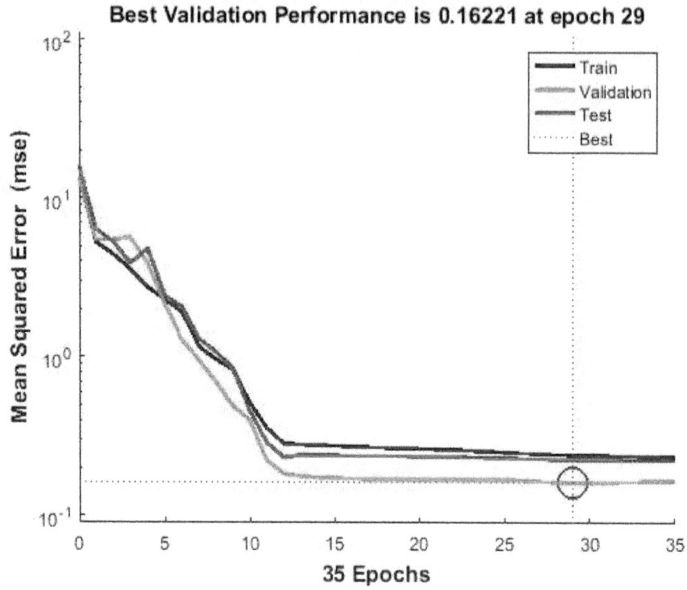


mean square errors for the best penetration network, these figures ensure there is no overfitting.

For height, best networks showed regression of 68 to 92 and MSE of 0.03 to 0.08. It consisted of 1 hidden layer containing 7 neurons of "tansig" function and an output layer with function of "purelin".

For better results it was suggested to make other networks for height and penetration related to the effective inputs known from the literature and table.1. None of the developed networks with single input single output showed reasonable results most of them showed over fitting and negative regression, so the multi input multi output are to be accepted. 'Traingd' function gave the best result for height. Fig.9 and fig.10 show regression and mean square errors for the best height network.

*Figure 7. Best penetration network regression.*

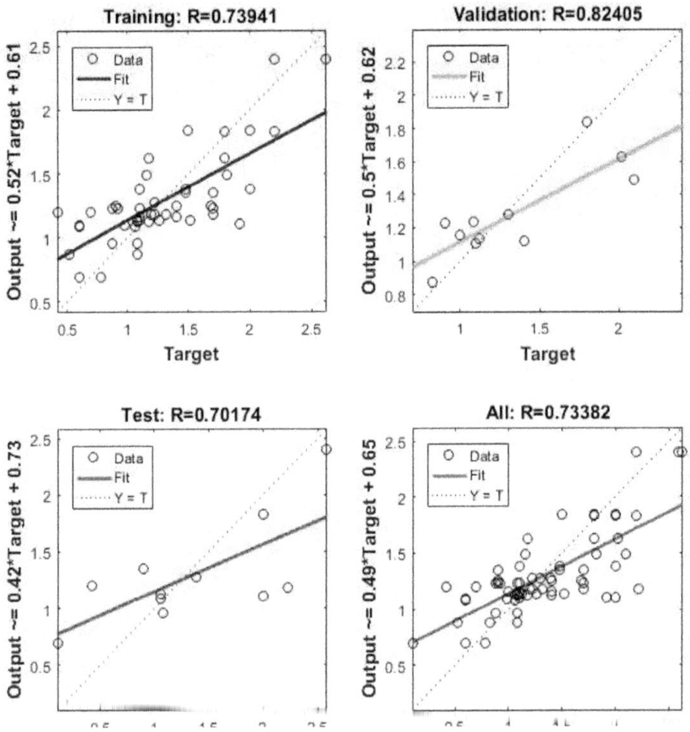


*Figure 8. Best penetration network MSE.*

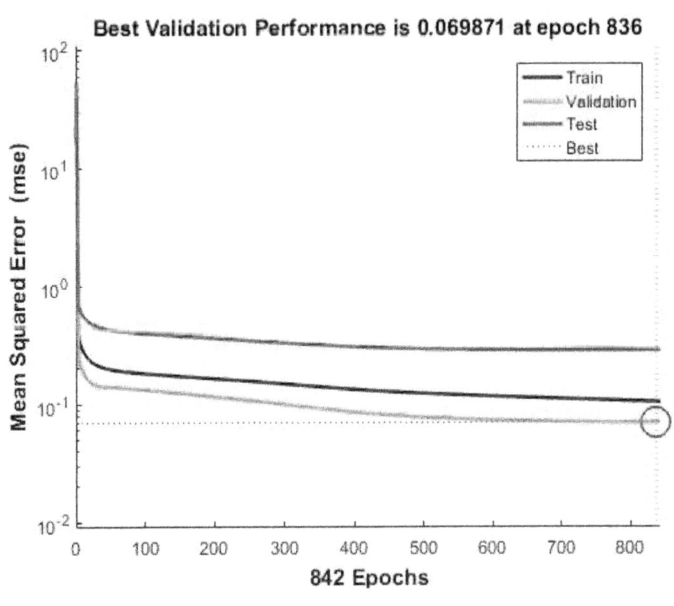

*Figure 9. Best height network regression.*

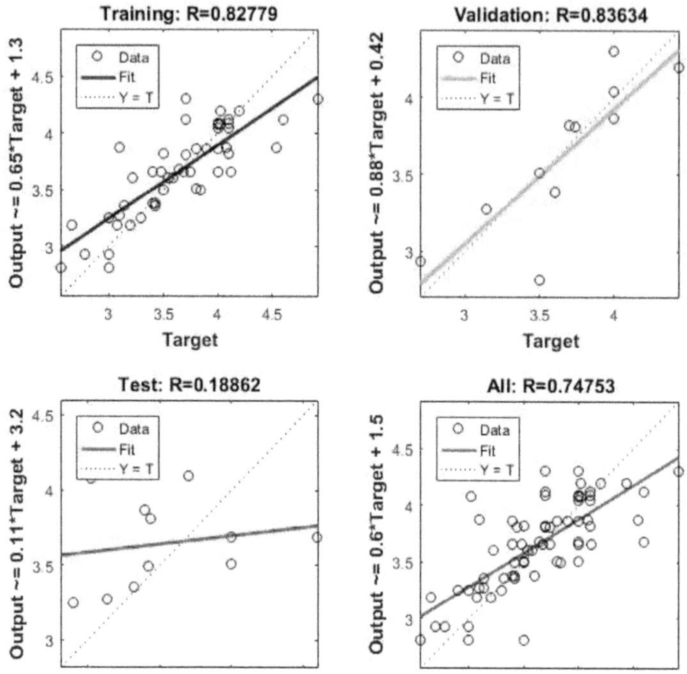


*Figure 10. Best height network MSE.*

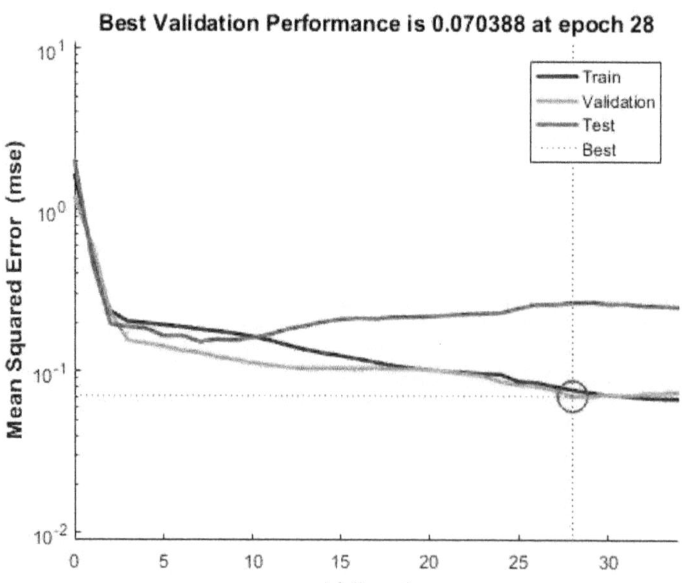

*Figure 11. relation between HAZ hardness and deposited line penetration.*

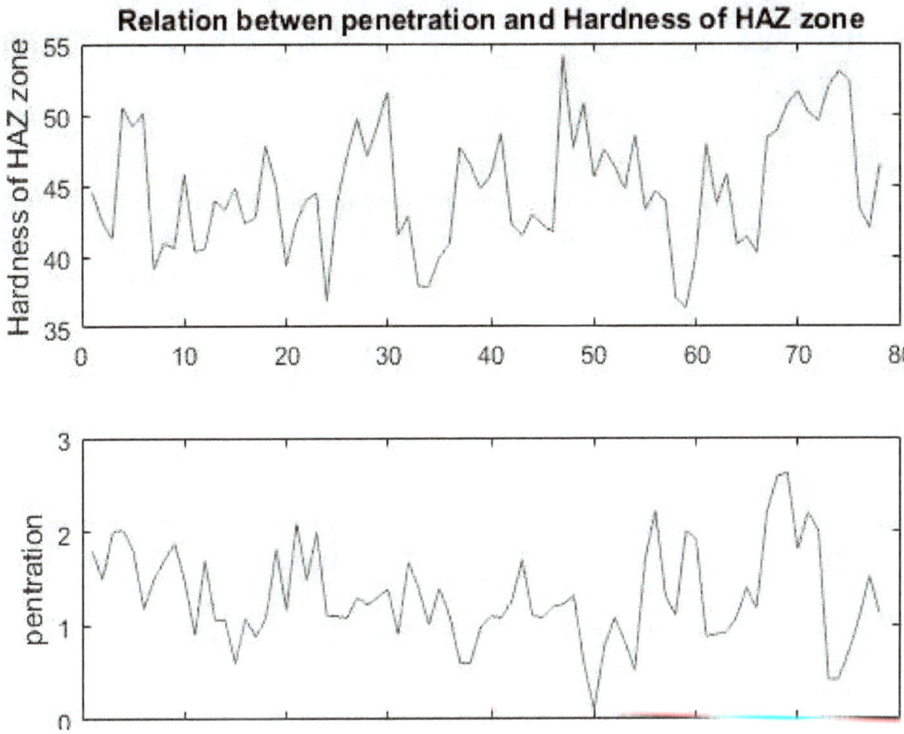


During taking measurements a relation between HAZ (heat affected zone) hardness and penetration of the laid bead was observed and modelling confirmed that, so penetration was used as a fourth input for HAZ hardness model. This model was used later to find the optimal parameters using different methods. Fig.11 shows this relation, this relation was not investigated in this work, but it is thought to be due to thermal dissipation factors.

## Optimization

Four optimization methodologies were used in this stage. One of them was developed as improvement for one of the other three ones. These methodologies are GA (genetic algorithm), ACO (ant colony optimization), HS (harmony search) and modified ACO (Ramadan et al., 2018). Modified ACO was developed merging HS with ACO to solve the problem of ACO initial parameters selection as will be discussed later

All optimization methodologies used the same objective function:

$$c.f = \left(1 - e^{-\beta 1}\right)\left(width\_mean - 1\right) + e^{-\beta 2}\left(deposited\_metal\_hardness - 49\right)\text{.........}(1)$$

Where β1 and β2 ranges from 0.1 to 1.

The optimization targets are the least mean width close enough to 1 mm the diameter of MIG wire and least deposited metal hardness close to 49 HRA highest hardness of the base metal. Also difference between deposited metal hardness and HAZ haedness must be taken in consideration to be positive and small as possible.

3 optimization techniques were tried individually then one was newly developed. Table.3 shows best results of the 3 techniques tried first.

*Table 3. optimization results comparison:*

| | width mean | deposited metal hardness | HAZ hardness | Penetration |
|---|---|---|---|---|
| Genetic | 5.458 | 60.46 | 59.58 | 0.5801 |
| Harmony | 4.606 | 61.69 | 60.9 | 1.245 |
| Ant | 5.167 | 59.96 | 58.6 | 1.191 |

According to table.3 harmony search technique showed least mean width, deposited metal hardness showed results close to each other. Taking in consideration the difference between deposited metal hardness and HAZ hardness harmony search will be the best, this parameter indicates the changes happening between different levels leading to better performance during accumulation.

Comparing the best results found by each technique shown in tables 4, 5 and 6 it can be noticed that:

1. Harmony search results are not repeated for different betas values, but they are very close to each other which mean this technique converged many times to a global minimum.
2. Genetic algorithm repeated each one of the best three solutions twice for different values of betas also they are very different from each other which means the technique did not converge properly a global minimum specially results from harmony are less than all of them.
3. Ant colony repeated one of the best results twice while another one (the best one) repeated 8 times. Comparing the best one with the results of harmony

*Table 4. Optimization results of harmony search*

| beta1 | beta2 | Voltage | Current | Travel speed | width mean | weld metal hardness | HAZ Hardness | penetration |
|---|---|---|---|---|---|---|---|---|
| 0.9 | 0.6 | 24.4364 | 181.6982 | 0.0127 | 5.038 | 62.08 | 61.82 | 1.532 |
| 0.9 | 0.1 | 24.5956 | 188.6754 | 0.0199 | 4.606 | 61.69 | 60.9 | 1.245 |
| 0.9 | 1 | 24.2983 | 183.4211 | 0.0104 | 5.022 | 61.24 | 59.92 | 1.326 |

*Table 5. Optimization results of genetic algorithm*

| beta1 | beta2 | Voltage | Current | Travel speed | width mean | weld metal hardness | HAZ Hardness | penetration |
|---|---|---|---|---|---|---|---|---|
| 0.4 | 0.2 | 24 | 185.2873 | 0.0249 | 8.432 | 59.67 | 59.01 | 1.028 |
| 0.9 | 0.6 | 24 | 185.2873 | 0.0249 | 8.432 | 59.67 | 59.01 | 1.028 |
| 0.5 | 0.5 | 24 | 188.2064 | 0.0049 | 5.458 | 60.46 | 59.58 | 0.5801 |

*Table 6. Optimization results of ant colony*

| beta1 | beta2 | Voltage | Current | Travel speed | width mean | weld metal hardness | HAZ Hardness | penetration |
|---|---|---|---|---|---|---|---|---|
| 0.8 | 0.1 | 24.7778 | 189.1919 | 0.0137 | 5.029 | 60.44 | 59.68 | 1.11 |
| 0.8 | 0.6 | 24.3737 | 182.0202 | 0.0072 | 5.199 | 60.38 | 59.26 | 1.496 |
| 0.8 | 0.6 | 24 | 186.0606 | 0.0239 | 5.199 | 60.38 | 59.26 | 1.496 |
| 0.1 | 0.4 | 24.3333 | 187.0707 | 0.0217 | 5.167 | 59.96 | 58.6 | 1.191 |
| 0.1 | 1 | 24.3333 | 187.0707 | 0.0217 | 5.167 | 59.96 | 58.6 | 1.191 |
| 0.2 | 0.4 | 24.3333 | 187.0707 | 0.0217 | 5.167 | 59.96 | 58.6 | 1.191 |
| 0.3 | 1 | 24.3333 | 187.0707 | 0.0217 | 5.167 | 59.96 | 58.6 | 1.191 |
| 0.4 | 0.4 | 24.3333 | 187.0707 | 0.0217 | 5.167 | 59.96 | 58.6 | 1.191 |
| 0.6 | 0.4 | 24.3333 | 187.0707 | 0.0217 | 5.167 | 59.96 | 58.6 | 1.191 |
| 0.9 | 0.6 | 24.3333 | 187.0707 | 0.0217 | 5.167 | 59.96 | 58.6 | 1.191 |

it is obvious results of harmony search are better. Also repetition of the best ant colony result although it is not the optimal and differences between ant colony results mean ant colony somehow was trapped in local minimum and did not converge properly to optimal solution.

For every methodology best result or results were practically verified. fig.12 shows deposition of one of the best results from genetic algorithm technique, fig.13 shows deposition of one of the best results from ant colony optimization and fig.14 shows deposition of one of the best results from harmony search technique.

*Figure 12. Verification specimen of genetic algorithm results.*

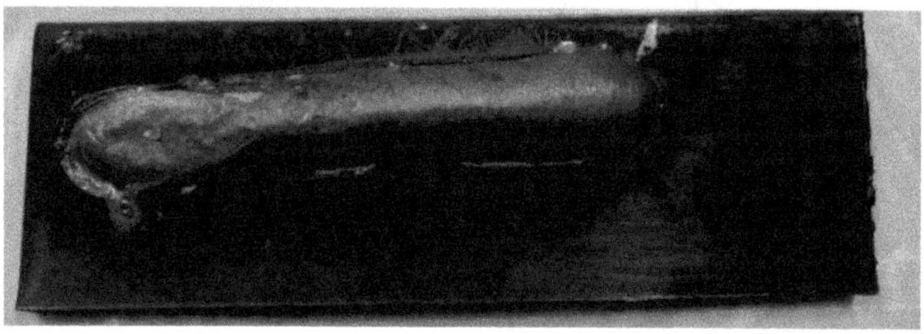

*Figure 13. Verification specimen of ant colony results.*

*Figure 14. Verification specimen of harmony search results.*

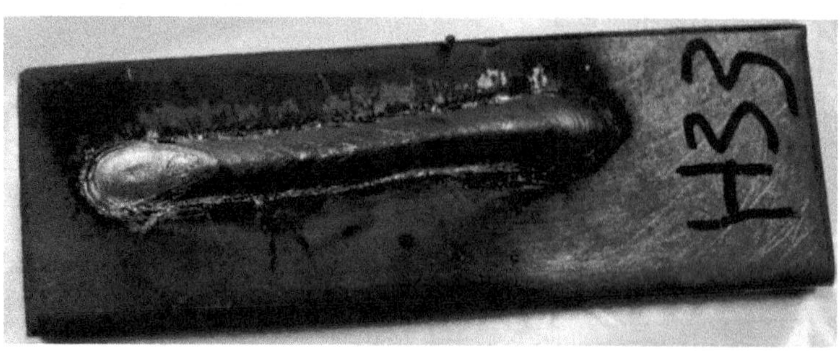

One of the main problems concerning Ant Colony optimization is determining initial values of ants, pheromone and evaporation. In order to modify performance of ant colony optimization these three values were selected by Harmony search optimization to get optimal values of them. To decrease number of trials best values of β1 and β2 shown in table.7 for both techniques were selected to be used together.

*Table 7. best values of β1 and β2:*

| Harmony search | | Ant colony | |
|---|---|---|---|
| β1 | β2 | β1 | β2 |
| 0.9 | 0.6 | 0.1 | 0.4 |
| 0.9 | 0.1 | 0.1 | 1 |
| 0.9 | 1 | 0.2 | 0.4 |
| | | 0.3 | 1 |
| | | 0.4 | 0.4 |
| | | 0.6 | 0.4 |
| | | 0.9 | 0.6 |
| | | 1 | 0.7 |
| | | 0.1 | 0.4 |

Also, here both used the same objective function. Harmony search was used to determine three parameters of ant colony optimization number of ants which ranged from 20 to 100 then reduced to 20 to 50 to decrease calculation time, pheromone which ranged from 0.1 to 1 and evaporation rate which ranged from

*Figure 15. How optimization works.*

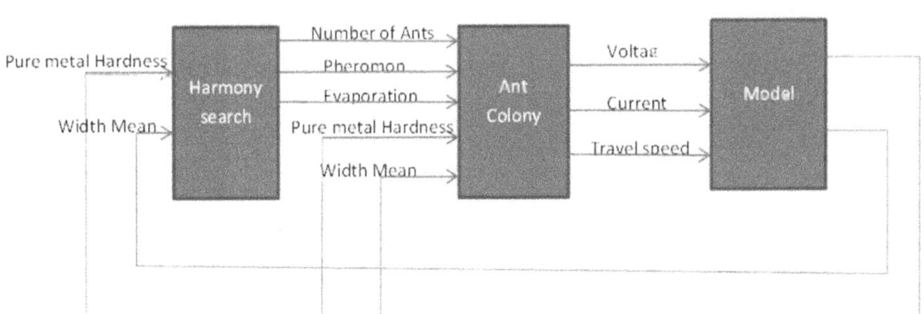

0.1 to 0.95. These values resulting from Harmony search optimization were sent to the ant colony optimization to use it finding the optimum values of process inputs satisfying the objective shown before, fig.15 shows how the optimization works.

Modified optimization steps are as follows:

1. Enter boundaries of the ACO parameters (ants, pheromone and evaporation) to the HS.
2. HS generates the vector of HM where every element of this vector contains 3 values one for each parameter of the ACO.
3. HS sends elements one by one to ACO to initialize it.
4. ACO starts working after receiving the initial values generating paths each one of them represents input set for the NN (neural network) model and continue processing as illustrated before.
5. When ACO finishes processing and select the fittest path (values) the corresponding NN output values are sent to the HS to evaluate fitness of the element it sent as initial parameters to ACO in step 3.
6. After testing all elements of the HM vector and ranking them according to their fitness, HS generates new elements of HM. each element is sent to ACO and its fitness is calculated then compared to fitness of the old HM. If the new element is better than one or more of the old HM vector it replaces the worst one.
7. HS continues generation and testing of elements till reaching the maximum number of trials. Noticing that in each trial it generates number of elements equal to length of the HM vector.
8. ACO runs number of times equal to (length of HM)* (number of HS trials).

Each two pairs were tested together 3 times. The best result was for values of $\beta 1= 0.9$, $\beta 2=0.1$ for harmony search and $\beta 1=0.1$, $\beta 2=0.4$ for ant colony. Table.8 shows comparison between Ant colony results and the one modified by Harmony search.

*Table 8. Comparison between Ant colony optimization results and Ant colony modified by Harmony search optimization results:*

| | Width mean | Pure metal hardness | Penetration |
|---|---|---|---|
| Ant | 5.167 | 59.96 | 1.191 |
| Ant modified by Harmony | 4.858 | 65.12 | 1.192 |

From table.8 it can be noticed that for width mean it decreased for the modified ant colony by 0.309 mm, pure metal hardness increased which agrees with notes at modeling stage although width was not approved as independent input.

Results of the modified ant colony technique was practically verified as shown in fig.16, table.9 shows comparison between best results from different techniques and practical verification.

*Figure 16. Verification specimen of modified ant colony results.*

*Table 9. Comparison between optimization simulated and practical results:*

| | | Voltage | Current | Travel speed | Mean width | penetration | Weld metal hardness | HAZ hardness |
|---|---|---|---|---|---|---|---|---|
| GA | simulated | 24 | 185.29 | 0.0249 | 8.432 | 1.028 | 59.67 | 59.01 |
| | practical | 24.1 | 188 | 0.0249 | 11.18 | 1.4 | 53.8 | 40 |
| ACO | simulated | 24.33 | 187.07 | 0.0217 | 5.167 | 1.191 | 59.96 | 58.6 |
| | practical | 24.3 | 187 | 0.0217 | 5.26 | 1.56 | 63.875 | 50.185 |
| MACO | simulated | 185 | 25 | 0.966 | 4.858 | 1.192 | 65.12 | 62.16 |
| | practical | 185 | 25 | 0.966 | 4.49 | 1.58 | 57.56 | 42.995 |
| HS | simulated | 24.596 | 188.68 | 0.0199 | 4.606 | 1.245 | 61.69 | 60.9 |
| | practical | 24.5 | 189 | 0.0199 | 4.36 | 1.17 | 59.37 | 47.28 |

As can be noticed in table.9 there is slight difference between simulation process inputs resulting from optimization and practical ones. This difference in voltage and current due to Fronious machine constrains while travel speed is always the same because it is gotten from robot motion.

From table.9 the best practical results were from modified ACO and HS but according to accumulation stage requirements HS results are slight better. HS results showed smaller width, smaller and not shallow penetration which mean there will be good mixing between layers without over melting for previous one each time, lower difference between weld metal hardness and HAZ hardness which decreases possibility of cracks in accumulation stage. Also all specimens showed homogenous width so it was not required to take standard deviation in consideration like the modelling stage.

## WAAM

Before starting WAAM (wire arc accumulation of metal) it was necessary to calibrate the robot to make sure it will be able to execute the horizontal and vertical offsets between lines and layers. From another research () the horizontal offset was known to be 0.738w where w is the single bead width. In this case the offset from the centre line of one line to the next one was 3.22 mm fig.17 show first layer deposition. It is nearly flat surface except for 2 notches due to temporary current dropping this photo was taken during a stop due to current drop then it was discarded and stated again for another one to get rid of cooling effect although it was for relatively short time.

For vertical offset trials with 0.3, 0.5 and 0.75 of the height were performed. 3 vertical lines were deposited one on another for each trial, but all of them were low. so trial with full height offset was performed before starting WAAM. This one with full height was the suitable one which indicate penetration is ideal and makes no over melting. 2 deposition samples were deposited each one consisted of 14 layers with 24 lines in each layer. The base metal was steel 37 blocks with dimensions of 200mm x 160 mm x 25 mm. Fig.18 shows one of the deposited samples. Robot was essential for deposition to achieve accurate offset between lines and layers. Also, Fronious machine attached to robot was the best for making sure voltage, current and feed rate are kept constant as possible during deposition unless electrical source current dropped.

6 tension specimens were cut from each WAAM specimen 3 of them parallel to deposited lines and the other 3 perpendicular to them. Table.10 summarises tension results. Only 4 specimens from each one was tested 2 from each direction. Only one of them was defected and included internal cavity so its result was excluded from calculating average results it is the marked one in table.10. Fig.19 shows one tension specimen after test and fig.20 shows stress strain diagram of one specimen.

Comparing values of yield and ultimate strength of the accumulated metal from table.4 with that of base metal (steel 37) according to DIN steel 37 is DIN 17100, its yield strength is 235 MPa and its ultimate strength ranges from 360 to 460 MPa, which mean accumulated specimens showed significantly higher yield strength and little higher ultimate strength. Also shape of specimen in fig.5 indicates it was a perfect ductile fracture. Which means layers and lines are sufficiently merged

*Figure 17. First deposited layer.*

*Figure 18. WAAM sample deposited by KUKA robot.*

*Table 10. Tension specimens results*

| specimen | Yield strength MPa | Ultimate strength MPa |
|---|---|---|
| 1st specimen 1st horizontal specimen | 360 | 480.64 |
| 1st specimen 2nd horizontal specimen | 340 | 453.74 |
| 1st specimen 1st vertical specimen | 365 | 485.97 |
| 1st specimen 2nd vertical specimen | 370 | 475.95 |
| 2nd specimen 2nd horizontal specimen | 360 | 479.25 |
| 2nd specimen 1st horizontal specimen | 360 | 477.25 |
| 2nd specimen 2nd vertical specimen | 360 | 483.34 |
| 2nd specimen 1st vertical specimen | 360 | 445.55 |
| average | 359.2857143 | 476.5914286 |

*Figure 19. Tension specimen.*

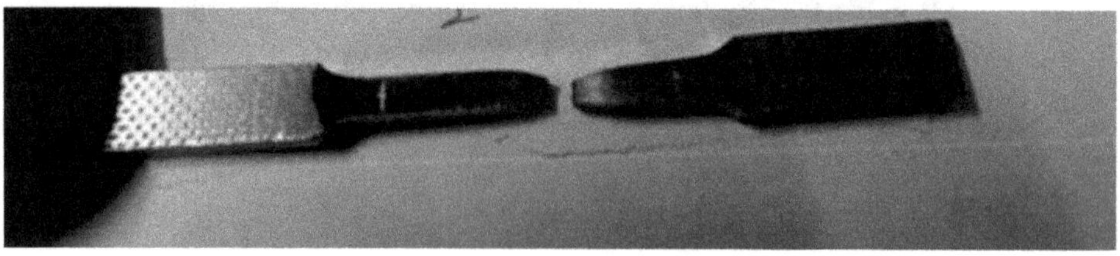

without over melting, lack of penetration or excessive heat affection. This indicates performance of robot with optimized MIG parameters was perfect for WAAM.

Comparing hardness of the deposited metal using optimized parameters from table.9 to the hardness of base-metal which was maximum 49 HRA, pure deposited metal hardness is 59.37 superficial hardness equivalents to 54.37 HRA and HAZ hardness is 47.28 HRA. Here small difference between pure metal hardness and base metal maximum hardness, while HAZ hardness is within limits of base metal hardness range and the difference between pure metal and HAZ hardness is small. This influenced the accumulated metal strength and made it optimal.

*Figure 20. Sample of the stress strain diagram of one specimen.*

| Description | | Result |
|---|---|---|
| Width x Thickness | W x T | 6.05 x 6.00 mm |
| Startsection | $S_0$ | 36.30 mm² |
| E-Modulus | E | 6.14 GPa |
| Load max | Fm | 16.47 kN |
| Rp at D1 | $R_{0.1}$ | 0.00 MPa |
| Rp at D2 | $R_{0.2}$ | 0.00 MPa |
| Tensile strength | $R_m$ | 453.74 MPa |
| Elongation | A | 44.67 % |
| Striction | Z | --- % |

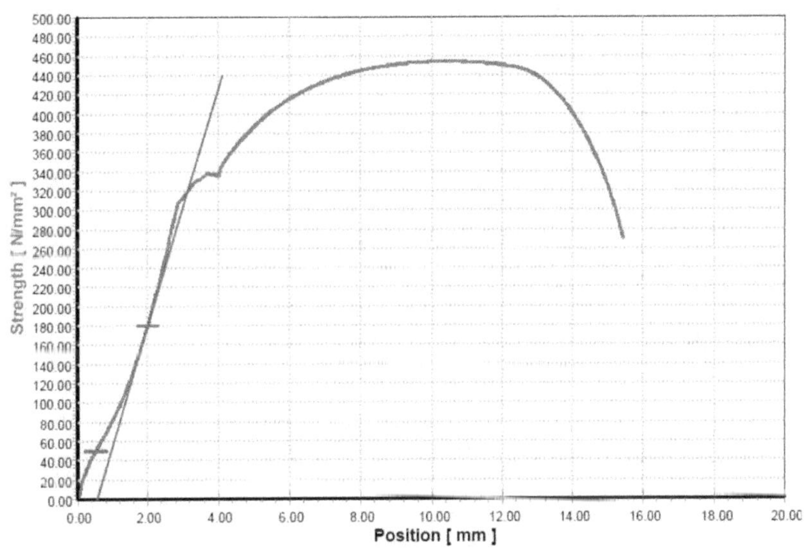


## FUTURE RESEARCH DIRECTIONS

- Apply MACO technique on swarm robots task planning to optimize them.
- Apply the technique on the same material with different thicknesses to make thickness an optimized parameter.
- Apply the technique on different materials with different thicknesses to make material and thickness as optimized parameters.

## CONCLUSION

- These results can be used for the same material in same conditions but for other materials the methodology must be repeated to get suitable values.
- Robot was essential to achieve good interference between lines and layers.
- Optimization and robot usage enabled accumulation of blocks with good mechanical properties without need for heat treatment. This makes it possible to combine WAAM with conventional machining, using WAAM for repair and building up while heat treatment is not available.
- Optimization with facilities offered by robot and its automatic MIG machine enabled decreasing line width to 4.36 mm with homogenous width and height this decreases post machining required after WAAM in different applications.
- HS was most suitable nature inspired optimization technique for this process while GA could not cope with process complication and nonlinearity.

## ACKNOWLEDGMENT

Thanks to NOZOM company and its' chairman for support.

## REFERENCES

Barakat, Abdel Ghany, Abd Elsalam, & El Atriby. (2018). Modeling of the Weld Bead Shape at Different Welding Parameters. *AMME-18 The 9th International Scientific Conference of The Military Technical College*, 3–5.

Cao, Y., Zhu, S., Liang, X., & Wang, W. (2011). Overlapping Model of Beads and Curve Fitting of Bead Section for Rapid Manufacturing by Robotic MAG Welding Process. *Robotics and Computer-integrated Manufacturing*, *27*(3), 641–645. doi:10.1016/j.rcim.2010.11.002

Cruz, J. G., Torres, E. M., & Sadek, C. (2015). A Methodology for Modeling and Control of Weld Bead Width in the GMAW Process. *Journal of the Brazilian Society of Mechanical Sciences and Engineering*, *37*(5), 1529–1541. doi:10.100740430-014-0299-8

Ding, D., Pan, Z., Cuiuri, D., Li, H., van Duin, S., & Larkin, N. (2016). Bead Modelling and Implementation of Adaptive MAT Path in Wire and Arc Additive Manufacturing. *Robotics and Computer-integrated Manufacturing*, *39*, 32–42. doi:10.1016/j.rcim.2015.12.004

Lostado-lorza, R., & Fernández-martínez, R. (2014). *International Joint Conference SOCO'13-CISIS'13-ICEUTE'13*. 10.1007/978-3-319-01854-6

Malviya, R., & Pratihar, D. K. (2011). Tuning of Neural Networks Using Particle Swarm Optimization to Model MIG Welding Process. *Swarm and Evolutionary Computation*, *1*(4), 223–235. doi:10.1016/j.swevo.2011.07.001

Nele, L., Sarno, E., & Keshari, A. (2013). Modeling of Multiple Characteristics of an Arc Weld Joint. *International Journal of Advanced Manufacturing Technology*, *69*(5–8), 1331–1341. doi:10.100700170-013-5077-8

Raghavendra, N., Koranne, R., Pal, S., Pal, S. K., & Samantaray, A. K. (2009). Joint Strength Prediction in a Pulsed MIG Welding Process Using Hybrid Neuro Ant Colony-Optimized Model. *International Journal of Advanced Manufacturing Technology*, *41*(7–8), 694–705. doi:10.100700170-008-1517-2

Ramadan, A., Barakat, A., & Elatriby, S. (2018). Optimizing Ant Colony Using Harmony Search Optimization. *4th International Conference Welding and Failure Analysis of Engineering, Materials*, 3–9. 10.13140/RG.2.2.10327.32161

Salonitis, K. (2016). Design for Additive Manufacturing Based on the Axiomatic Design Method. *International Journal of Advanced Manufacturing Technology*, *87*(1–4), 989–996. doi:10.100700170-016-8540-5

Sumesh, A., Rameshkumar, K., Raja, A., Mohandas, K., Santhakumari, A., & Shyambabu, R. (2017). Establishing Correlation Between Current and Voltage Signatures of the Arc and Weld Defects in GMAW Process. *Arabian Journal for Science and Engineering*, *42*(11), 4649–4665. doi:10.100713369-017-2609-9

Xiong, J., Zhang, G., Hu, J., & Li, Y. (2013). Forecasting Process Parameters for GMAW-Based Rapid Manufacturing Using Closed-Loop Iteration Based on Neural Network. *International Journal of Advanced Manufacturing Technology*, *69*(1–4), 743–751. doi:10.100700170-013-5038-2

# Chapter 7
# Additive Manufacturing Feature Taxonomy and Placement of Parts in AM Enclosure

**Prafull Agarwal**
*CAD Lab, Department of Mechanical Engineering, Manipal University Jaipur, India*

**Rishi Kurian**
*CAD Lab, Department of Mechanical Engineering, Manipal University Jaipur, India*

**Ravi Kumar Gupta**
https://orcid.org/0000-0003-4419-8111
*Manipal Institute of Technology Bengaluru, Manipal Academy of Higher Education, Manipal, India*

## ABSTRACT

*Additive Manufacturing (AM) is a layer-by-layer deposition of material for the production of the desired product. The design flexibility associated with AM is much more when compared to the conventional manufacturing process. To manufacture a part with AM, two things play a critical role: the designing of the part and the other is the placement of the part in the build volume. As already mentioned, design flexibility associated with AM is much more when compared to the conventional manufacturing process. However, to correctly implement the design flexibility, we need a knowledge base at our disposal so that appropriate features can be used for the part production. The AM feature taxonomy forms the backbone of the knowledge base. The taxonomy comprises AM features classified based on different categories, which helps us understand every feature's*

DOI: 10.4018/978-1-7998-8516-0.ch007

*importance. Talking about the part placement, we know that optimal placement is the key factor that makes the AM process economically feasible.*

## INTRODUCTION

Additive Manufacturing (AM) refers to the automated production of parts from their CAD model. In AM, the desired physical model is achieved with the help of layer-by-layer construction of raw material as required form the CAD model by a controlled material-additive process. Rather than AM, the subtractive manufacturing is a process in which raw material is cut to a desired final shape and size by a controlled material-removal process. The advantages with AM are immense, including the drastic decrease in the production time, reduction of wastage, assembly can be produced as single part. AM process has been primarily evolved from the Rapid Prototyping process, which in their holistic sense confer to the Rapid Production of parts. Parts produced with AM are thus possible to be made in hours and not months or weeks, which is primarily associated with conventional production methods. There are different phrasings utilized at present that could be viewed as options in contrast to the globally perceived term of AM, like 3D Printing, additive fabrication, layered manufacturing, layered free-framing, freeform manufacture, layer-based manufacturing, and rapid manufacturing. These wording varieties have come about because of different points of view that show up from the industry and academics that utilize the systems. For convenience, we will be making use of the term AM and 3D Printing throughout this report.

### 1.1 Historical perspective of Additive Manufacturing

The advances in AM, popularly known as 3D Printing, have opened the doors for some highly significant innovations in almost every perceivable field. 3D Printing refers to the art of producing a wide range of complex geometries from their 3D CAD models. The discovery of 3D Printing can be credited to Charles Hull in the early 1980s, with it being initially called Stereolithography (Schubert et al., 2014). The process involved printing layers of work material one over the other to get the desired 3D object. Over a while, the process of 3D Printing has evolved tremendously and now many more techniques such as Fused Deposition Modelling, Powder Bed Fusion and Inkjet Printing are regularly used to produce complex geometries and objects (Bandyopadhyay et al., 2015). The process of 3D Printing has been using in various fields such as construction and manufacturing. Also, it finds an application in rapid prototyping, which is the rapid production of the model, faster than the normal process (Ngo et al., 2018).

## 1.2 Applications of Additive Manufacturing

With the developments taking place, researchers are constantly looking for new avenues where the 3D printing process can produce a significant difference. One of the key advantages of 3D Printing is the accuracy and precision associated with the process (Mandrycky et al., 2016). Since the entire production process is automated, the models produced are accurate and precise, involving minor human errors. Apart from this, the repeatability associated with the process is very high, which means we can produce the same model repeatedly while maintaining the same level of accuracy and precision. Over a while, 3D printing technology has taken an upward curve, and the primary driving force for that increase is the reduced cost and enhanced efficiency of the 3D printers, which means the increased accessibility to the masses. The initial use of the process was limited to producing models by engineers and architects until very recently, when it is used in mass production began (Berman, 2012). 3D Printing offers us the unique property of mass customization wherein each subsequent product is different from the other without incurring any increase in the cost because mass production is still taking place (Ngo et al., 2018). All this while the cost of the first item and the last item remaining the same, providing us with a better approximation of the costs associated with the entire process.

Due to the advantage associated with 3D Printing, various manufacturing industries can incorporate their use. The most prominent users of the technology include the automotive and aerospace industry (Lim et al., 2016). The efficiency of the process in prototyping is what primarily promotes its use in the industries. Apart from that, as discussed above, the ability to produce intricate, complex geometries while at the same time being able to perform modifications on the design without having to incur any additional costs make the use of the process an almost game changer. Other industries that have benefited immensely from the developments in the process include the jewellery industry and medical (Murphy & Atala, 2014; Yap & Yeong, 2014).

## 1.3 Additive Manufacturing Processes

Additive Manufacturing also referred as 3D printing can be divided into two classes, first one is based on the physical state of the raw material used, i.e., liquid-, solid- or powder-based processes and other one is based on the manner in which the matter is fused at a molecular level, i.e., thermal, ultraviolet light, laser, or electrons beam. The classifications of AM based on material used and then further based on the various processes are presented in Figure 1.

*Figure 1. Classification of AM processes [9]*

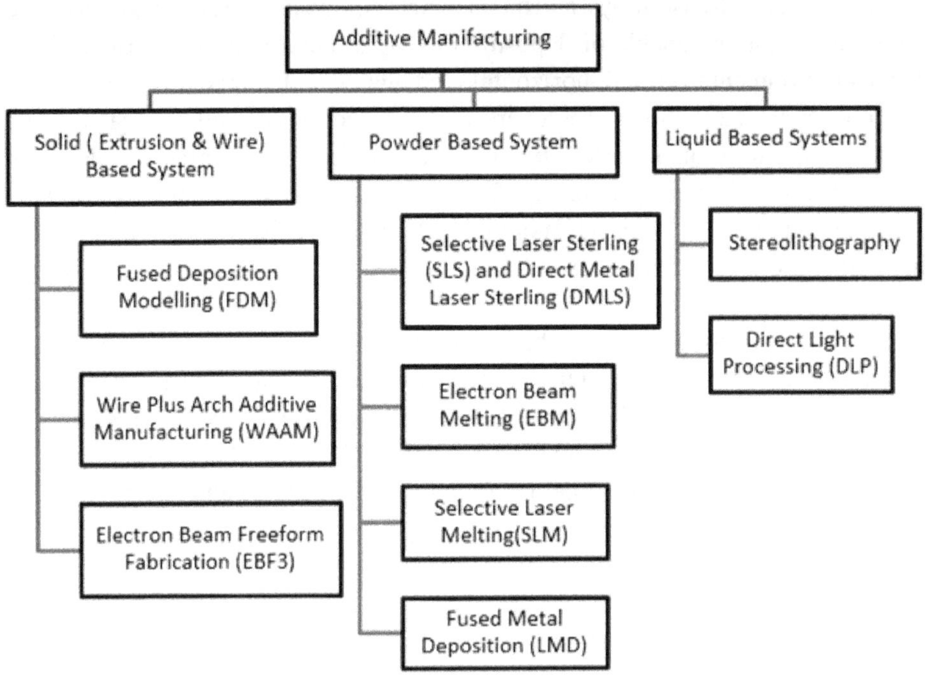


Selective Laser Sintering (SLS), Selective Laser Melting (SLM), Direct Metal Laser Sintering (DMLS), Electron Beam Melting (EBM), and Wire plus Arc Additive Manufacturing (WAAM) are just a few of the numerous AM processes (Figure 1). These processes can create amazingly dense parts whose mechanical and electrochemical properties that remain at standard with those delivered by other ordinary manufacturing techniques without the requirement for post-preparing. SLS and DMLS are essentially identical processes. While SLS can be utilized to manufacture parts made of an assortment of materials like plastics, earthenware production, and metals, DMLS can only be used to fabricate parts made of metal alloys. SLM can fabricate any material. It heats and melts the powdered material using a high-energy laser beam. EBM processes only metals because it utilizes a high-energy density electron beam which can produce dense and solid parts. It has currently found itself as a high-quality alternative to laser melting when it comes to manufacturing and repairing turbine blades.

Friction Stir Additive Manufacturing (FSAM) is another additive manufacturing procedure that not only improves mechanical properties, but also produces properties that are distinct from those obtained through conventional methods. Another method is electron beam freeform fabrication (EBF3), which can be used to fabricate structures made of aluminium, high strength steels, titanium, titanium aluminides, nickel-base alloys, and metal matrix composites. WAAM is alternative wire-based system capable of deposition of weld metal in an unrestricted free-space environment. Vertical, horizontal, and angled walls, mixed-material conic sections, enclosed sections, crossovers, and intersections can all be created using this process. Additionally, because it is not constrained by a cabinet, it allows for the production of larger components. The process utilizes commercially available aerospace-grade wire. The products attained through WAAM are exceptionally high in quality and are even better than those obtained by the alternative conservative welding procedures.

AM or 3D Printing is a big umbrella that encompasses various other processes to achieve the desired result. Of the various processes available at our disposal, some of them including, Stereolithography (STL), Fused Deposition Modelling (FDM), Laser Sintering (LS) and Material Jetting (MJ), have been discussed here. These are easily accessible as being commercially available and inexpensive.

### 1.3.1 Stereolithography

Stereolithography is a 3D printing process that uses a photochemical process to create polymers by crosslinking monomers and oligomers. It has high printing quality and speed. UV light source, which is the conventional polymerization method used, can cause cancer and are reportedly harmful to DNA cells, due to which we prefer visible light Stereolithography (Hull, 1984; Sinha & Häder, 2002). Chuck Hull gave the term 'stereolithography' in his patent filed in 1984 (Hull, 1984). It has undergone many developments over the years resulting in a better surface finish and quality. These advancements have resulted in adapting this method to various industries, especially in the biomedical industry (Melchels et al., 2010). Stereolithography has more fabrication accuracy and resolution compared to other additive layer methods (Pham & Gault, 1998).

The stereolithography process is based on the photo-polymerization of monomers and oligomers to fabricate 3D structures based on computer scans. We can classify them into two categories as Single photon, and Multi-photon (Bártolo, 2011).

The first step is creating a 3D model of the part using computer-aided design software by designing the model of using advanced scanners in respect to biomedical application, many medical scanning equipment such as CT scan and MRI aid in creating the intricate parts of human anatomy. After that, we translate the 3D models to standard tessellation language format (STL) and special software used to slice the model to layers of specified thickness and transferring the data to the apparatus (Raman & Bashir, 2015). Figure 2 explains the process flow of stereolithography fabrication on 3D models.

*Figure 2. Process flow of stereolithography fabrication on 3D models*

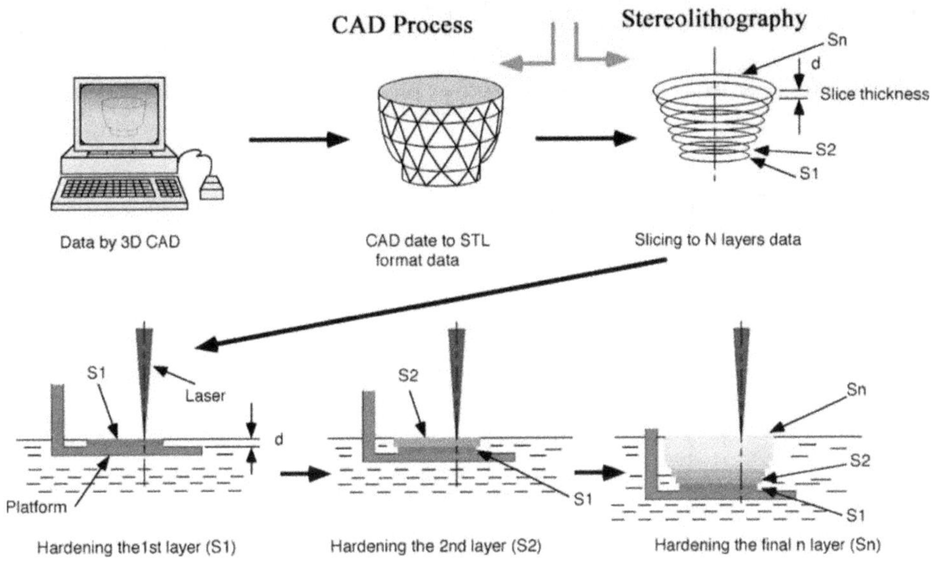


### *1.3.1.1 Single Photon Stereolithography*

The absorption of a single photon causes photopolymerization in single-photon stereolithography, as the name implies. This technology encompasses UV stereolithography and visible stereolithography. Direct laser writing employs a high-energy beam to trace a 2-dimensional cross-section on a layer of resin and then produces a 3-dimensional model by sequentially polymerizing a layer at a time from the bottom up (Bártolo, 2011). The mask based Stereolithography creates the entire 2-D cross-section in single exposure rather than tracing like in direct laser writing apparatus; therefore, it has faster output, thereby saving time. There is a mismatch between part size and resolution in this apparatus. Combining laser light and mask-based methods, which are not in mainstream use yet, help us immensely in tackling these problems, which are yet to be widely used (Johnson et al., 2013).

### *1.3.1.2 Multi-Photon Stereolithography*

Multiphoton Stereolithography consists of sequential or simultaneous absorption of two low-intensity photons for photo-polymerization of the resin. In two-photon Stereolithography, the liquid resin is fired by light waves from an infrared laser. The use of multiphoton SLA to create sophisticated interior microarchitectures for tissue engineering holds a lot of promise (Weiß, 2009).

*Figure 3. (A) Mechanism of TPA when simultaneous excitation occurs. (B) spatial compression using high numerical aperture, (C) temporal compression using ultrafast lasers (Spangenberg, 2013)*

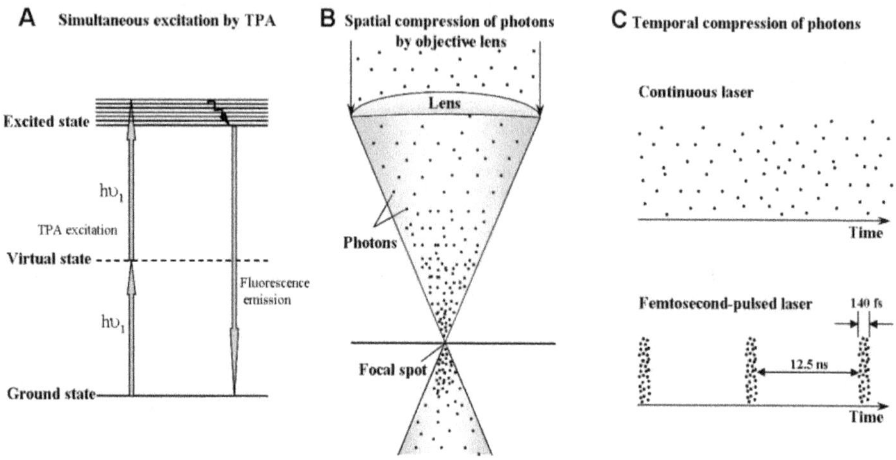


## 1.3.2 Fused Deposition Modelling

Fused deposition modelling, frequently abbreviated as FDM, was created, and is only marketed by Stratasys Inc. FDM is distinguished from SLA and LS systems by the absence of a laser used to construct the part's layers. The FDM machine operates on the same basis as a three-axis numerical control machine tool. As illustrated in Figure 3, a nozzle is controlled by a computer along three axes and extrudes the explicit material that has been melted via heating. The material exits the nozzle in a semi-liquid state and instantly hardens to the ambient temperature. As a result, temperature balancing of the liquid modelling material just above the solidification point is critical for the FDM process. The FDM head is fed by a reel of modelling material filament that is wound around the head. A transformation from one substance to another can be accomplished very quickly (Ziemian & Crawn, 2001).

Numerous materials are present having varying resistance-temperature trade-offs. The FDM process may also be employed with polycarbonates, polyphenylsulfones and waxes (Chua et al., 2010). Usually, processing standard FDM materials leads to lowering in densities when we compare it to components produced using injection moulding. Infiltration as post- processing is generally used to boost density. Figure 4 shows an example of a bike gauge pod constructed of FDM polycarbonate material. Like conventional investment casting wax, FDM components may also be treated without changing the standard investment casting process.

*Figure 4. Fused Deposition Modelling Process*

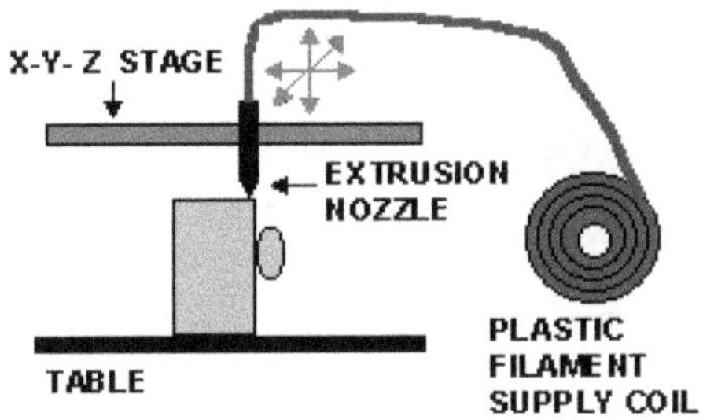


FDM, like all other 3D printing processes, has its drawbacks. Although it is simple to make clean components with FDM, it is not the most precise technique. Furthermore, the printing process leaves a visible player in 3D printed things, necessitating some post-processing to provide a smooth surface and erase the layers of the object.

### 1.3.3 Selective Laser Sintering

Selective laser sintering (SLS) is one of the technologies that modern 3D printers employ. SLS is a process in which microscopic particles of plastic, ceramic, or glass are fused by the heat generated by a high-powered laser to form a solid, three-dimensional item (Figure 5). Carl Deckard, undergraduate student at the University of Texas and his mechanical engineering professor, Joe Beaman invented and patented the SLS method in the 1980s. In 1989, Deckard and Beaman co-founded one of the earliest 3D printing firms, Desk Top Manufacturing (DTM) Corp. 3D Systems, acquired DTM in 2001, a company that had already invented its own, but very different, the technology of 3D Printing called Stereolithography. In comparison to other AM technologies, laser sintering can make parts from a reasonably wide variety of commercially accessible powder materials, including polymers (nylon, glass-filled nylon, and polystyrene), metals (steel, titanium, alloy combinations, and composites), and foundry sand (Kruth, 2003). Complete melting, partial melting, or liquid-phase sintering are all possible physical processes. Up to 100% density can be reached with material qualities equal to those obtained using conventional production methods, depending on the material (Kruth, 2007). Generally, a massive number of parts can be packed within the powder bed, resulting in extremely high productivity (Wohlers, 2009).

*Figure 5. Selective Laser Sintering Process*

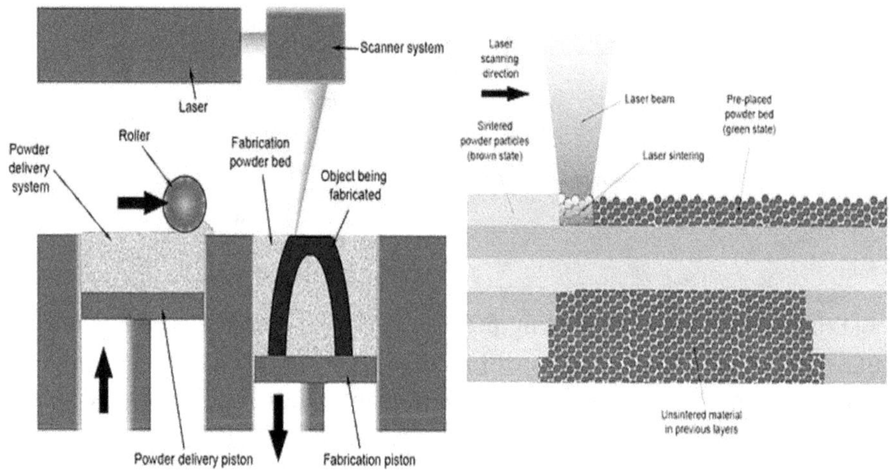


While Deckard and Beaman were the first to employ selective laser sintering to make a three-dimensional object, but they were far from the first to use sintering the process of producing objects from powders through atomic diffusion. Sintering has been used to make commonplace items such as bricks, porcelain, and jewellery for thousands of years. An object manufactured with an SLS machine begins as a computer-aided design (CAD) file, just like any other technique of 3D Printing. CAD files are transformed to the. STL format, which is recognized by 3D printers. Powder materials, most frequently polymers such as nylon, are scattered in a thin layer on top of the build platform within an SLS machine to produce objects produced with SLS. On the platform, a laser flashes forth, sketching an outline of the material onto the powder, which is fully automated by the means of a computer. The laser heats the powder to somewhere below (sintering) a little over (melting) its boiling temperature, uniting the powder's granules to form a solid. After the initial layer is formed, the SLS machine's base drops little less than 0.1mm, revealing a fresh powder layer for the laser to retrace and merge. Until the entire thing has been produced, this procedure is repeated. The properly developed product is left to cool before being withdrawn from the machine. SLS, unlike other 3D printing technologies, requires very little further processing after an object has been produced, which means that things usually need not be smoothed or otherwise tweaked once they have been removed from the SLS machine. Additionally, the use of external supports for holding the objects in place is minimal in SLS. Other 3D printing techniques, such as stereolithography or fused deposition modelling, frequently require similar supports, which makes them take longer than SLS.

LS technology is widely utilised around the world due to its ability to swiftly construct very complex shapes straight from CAD models (Wohlers, 2009). What initially began as a method for producing prototypes early in the designing process, it is gradually being used to generate end-use parts in small run production. (Rochus, 2007; Yadroitsev, 2007; Levy et al., 2003).

### 1.3.4 Material Jetting

Material Jetting (MJ) is an additive manufacturing (AM) technology which works in a similar way to two-dimensional printing. A printhead (similar to those seen in normal inkjet printing) discharges drops of a photosensitive material that is hardened using ultraviolet (UV) light, separating the layers layer by layer. The liquid polymers (acrylics) used in MJ are thermoset photopolymers. MJ 3D Printing creates components that are dimensionally precise and have a perfect surface

*Figure 6. Coloured MJ printer (Cheng & Huang, 2020*

quality. Material Jetting provides multi-material printing as well as a wide range of materials (which include materials similar to ABS, Rubber, and Transparent Material). Because of the above capabilities, MJ is a proven method for making prototypes and manufacturing tools. However, material jetting has a number of severe drawbacks, which we will address here. In one variant of the MJ process, drop-on-demand (DOD) printheads are utilised to spread viscous liquids and make wax-like parts. DOD primarily is almost only to make investment casting patterns. An example of a coloured MJ Printer is shown in Figure 6.

In contrast to most other 3D printing processes, MJ deposits material in a line-by-line method. Multiple inkjet printheads are mounted side by side on the same carrier, depositing material across the entire print surface in a single pass. This enables the employment of multiple heads to dispense distinct materials, making multi-material Printing, full-colour Printing, and dispensing of dissolvable support structures widespread and straightforward. Support structures are needed for material jetting and are later removed during the post-processing phase. The liquid substance solidifies in Material Jetting using a process called photo-polymerization. This is very much identical to the SLA process. Like SLA, material jetted items have uniform mechanical and thermal properties, however, unlike SLA, they do not require further post-curing to reach their optimal qualities due to the minimal layer height used.

With the primary discussion, on AM done, we move onto the discussion of the important avenue i.e., AM features. In the next section we review the taxonomies which have been previously prevented by various other researchers and scholars.

## 2. LITERATURE REVIEW ON SHAPE FEATURES

CAD (computer-aided design) technologies were popular in the 1970s. Professional designers used these tools to create a digital model of their goods. Although CAD systems improved design productivity, they lacked one crucial component: a connection to computer-aided production. Products could be created more efficiently and accurately if there was a decent means to communicate between the two, but a significant portion of the design had to be recreated throughout the production process (Gupta & Gurumoorthy, 2021; Gupta & Gurumoorthy, 2008; Nalluri, 1994). This was the driving force for the creation of features. Despite the lack of a comprehensive description in the literature, features are commonly defined as “particular sections of a shape to which particular information may be assigned”. This description is intentionally ambiguous, and certain information was first misinterpreted as ‘manufacturing information’. In the 1980s, methods for extracting manufacturing data from CAD models were developed by recognizing characteristics in the input data. Graph-based searching, convex hull decomposition, and hint-based approaches were created as feature recognition approaches.

Features are commonly employed in product modelling and play a crucial part in CAD/CAM (computer aided design and manufacturing), including design and production information. Design by features, also known as feature-based modelling, is a product design methodology that allows designers to develop a product model by selecting characteristics from a library (Fontana et al., 1999). Attaching, inserting, removing, positioning, and modifying the dimension of a feature into a model are just a few of the editing procedures available. In this methodology, we tend to involve the features right at the start of the modelling process.

The two categories of form features in design are regular shaped and freeform features (Nyirenda & Bronsvoort, 2009). Slots, holes, pockets, and ribs, which are regularly created features in spherical, prismatic, or cylindrical geometries, are widely used in CAD modelling. Freeform features can be classified as either freeform volume or freeform surface characteristics (Pernot et al., 2008). Solid components are modelled with volume characteristics, whereas sheet products are modelled using surface features.

The primary function is either of the two things: to achieve the desired function or to give a specific appearance to a part. Keeping that in mind, it is also possible to classify the form features according to the application of the feature in part. A feature taxonomy based on static and kinetic traits was presented by (Cunningham & Dixon, 1988). Static features were classified into the following five groups based on their functional intent:

a. Primitive (major shape);
b. As an add-on (local modification);
c. Intersections (basic add-on interaction) ;
d. Full form (attributes for the entire part);
e. Macros (combinations of primitives).

Pratt & Wilson (Wilson & Pratt, 1988) proposed a classification of solid model characteristics for use in a solid modelling system as depicted in Figure 7.

With regard to the classification of free form features, the work of Fontana et al. (Fontana et al., 1999) and Nyirenda et al. (Nyirenda & Bronsvoort, 2009) is extremely important. Fontana et al. (1999), on the one hand, investigated the characteristics of products with extremely complex shapes and geometries that we can identify and recognize. Whereas Nyirenda et al. (2009), on the other hand, classified the characteristics of products based on shape characteristics and their relative topologies. These freeform features are divided into four categories according to this taxonomy: deform, cut, transition, and compound as presented in Figure 8.

*Figure 7. Pratt and Wilson's Solid Model Features Taxonomy (Wilson & Pratt, 1988)*

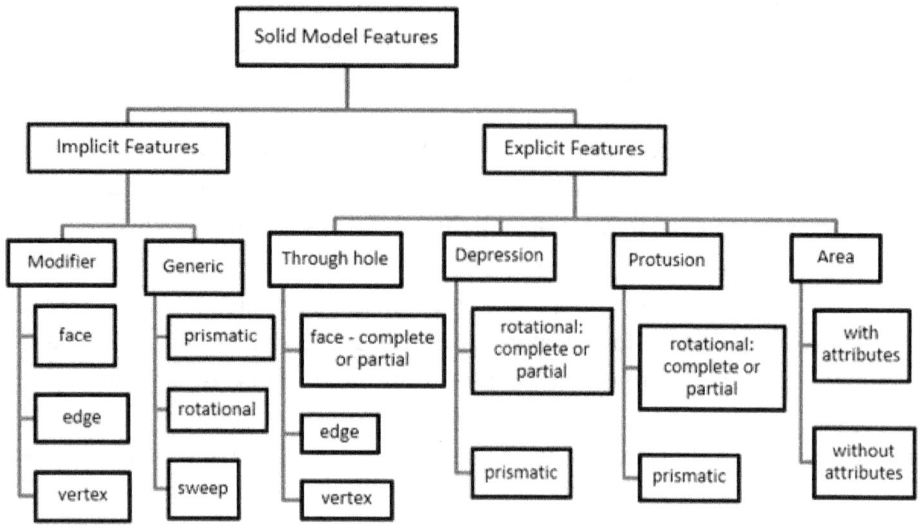


Deform features are protrusions in the model that represent material displacement and are represented by deform features. When a cut feature is present in the model, it represents material that has been removed from the model. Transition features, such as a blend, are characteristics that connect two or more different things. They can be additive or subtractive in nature. Any two or more fundamental features can be combined to form a compound feature, and vice versa (i.e., deform, cut or transition). The work of Fontana et al. is particularly significant because it served as a springboard for subsequent research into additive manufacturing while also depicting the classification of features that are recognizable for complex shapes and geometries.

The product taxonomy developed by Gupta & Gurumoorthy in 2013 (Gupta & Gurumoorthy, 2013) classifies the features on the basis of volumetric features, deformation features, and free form features as presented in Figure 9. The taxonomy is important as it enables us to distinguish the features on the basis of application in design. Subsequently, this taxonomy is extended for the knowledge-based development for additive manufacturing (Zhang, 2016) and semantic interoperability in product development (Gupta & Gurumoorthy, 2021). A lot of these features have been implemented with software.

The study of these taxonomies helps us in the proper understanding of the concept of how taxonomies are developed. With that in mind we move onto the need and development of a new taxonomy for products to be manufactured by AM process.

*Figure 8. Generic Free Form Taxonomy by Nyirenda et al. (Nyirenda & Bronsvoort, 2009)*

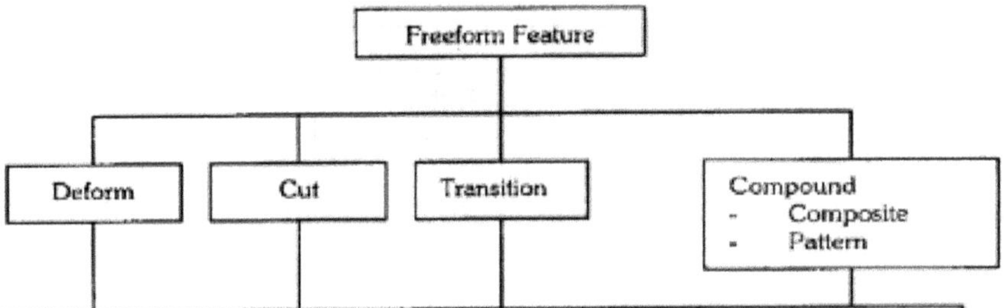


| | Deform | Cut | Transition | Compound |
|---|---|---|---|---|
| Border | Examples<br>- *ridge (open one end)*<br>- *groove (open one end)*<br>- *step*<br>- *mixed*<br>- … | Examples<br>- *end-cut (open sharp cut)*<br>- *slit(open sharp cut)*<br>- *notch*<br>- *mixed*<br>- … | Examples<br>- *extension (open join)*<br>- *boundary fill (on cut)*<br>- *mixed*<br>… | Examples<br>- *ridge with end-cut*<br>- … |
| Internal | Examples<br>- *ridge (internal)*<br>- *groove (internal)*<br>- *bump*<br>- *trough (internal)*<br>- *mixed*<br>- … | Examples<br>- *end-cut (sharp cut)*<br>- *hole (perforation)*<br>- *mixed*<br>… | Examples<br>- *patch-up (fill)*<br>- *fairing (fill-smoothening)*<br>- *fillet (internal)*<br>- *blend (internal)*<br>- *mixed*<br>… | Examples<br>- *hole on blend*<br>- … |
| Channel | Examples<br>- *ridge (open ends)*<br>- *groove (open ends)*<br>- *mixed*<br>- … | Examples<br>- *gap (through cut)*<br>- *mixed*<br>… | Examples<br>- *fillet (open ends)*<br>- *blend (open ends)*<br>- *mixed*<br>… | Examples<br>- *gap on groove*<br>- … |

*Figure 9. Unified shape Feature Taxonomy [28]*

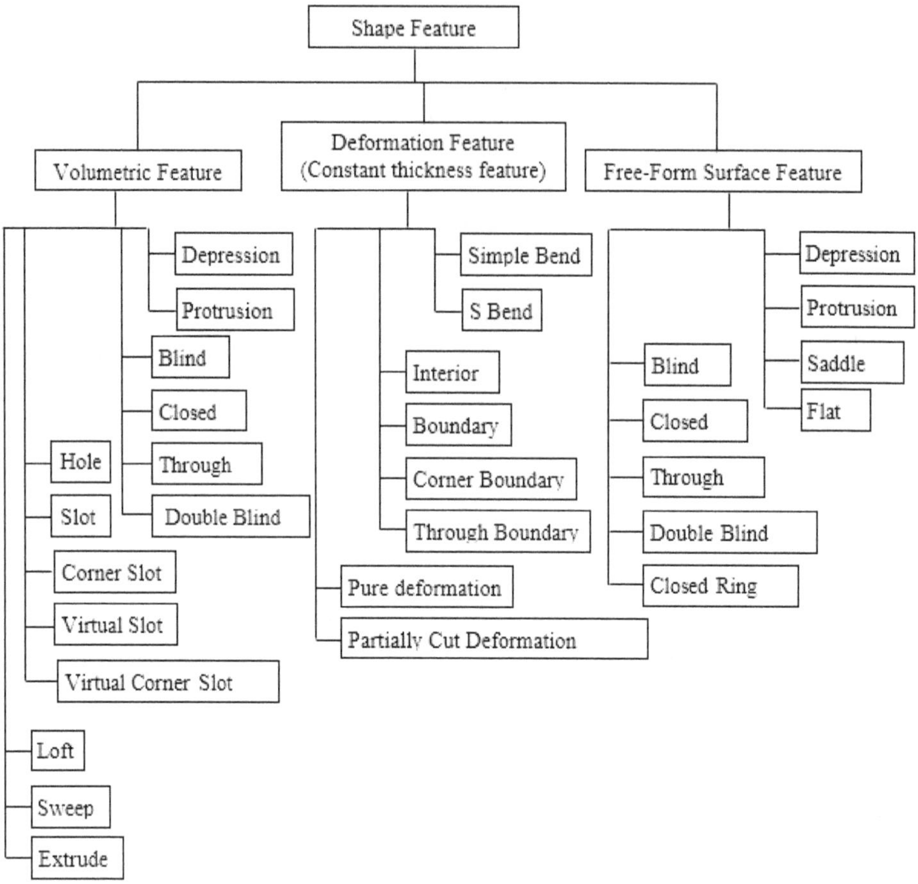


## 2.1 Need for Feature Taxonomy for AM

The work of the researchers has provided us with sufficient knowledge about the numerous feature classifications for AM, as well as a solid foundation upon which to build our further research. When we take a deeper look at the additive manufacturing process, we see that it is not just about the shape features, but also about what we can accomplish with the AM process as well. Using the taxonomies described above as a starting point, we may identify elements that are unique to additive manufacturing in terms of shape and functionality when compared to conventional manufacturing procedures, and we can classify them according to different parameters.

## 3. DEVELOPMENT OF FEATURE TAXONOMY for AM

Talking about taxonomies, De Candolle, a French botanist, coined the term taxonomy, which is derived from the Greek words taxis (order) and nomos (study) (Ostergaard & Summers, 2009). Thus, taxonomy is the study of configurations. Taxonomy is regarded as a branch of theory; it is a system for organizing complicated phenomena to facilitate comparison (Shneiderman & Norman, 1992). Researchers have used taxonomy for a variety of purposes, including information explanations (such as the taxonomy of social purposes in public schools (Derr, 1973), information exchange (such as an introduction to plant taxonomy), and the organisation of huge amounts of info (such as that generated during the elicitation of customer needs (Morris & Stauffer, 1994). Taxonomy is the classification and division of species and subgroups based on their common predecessors.

Mechanical design methods, tools, and hypotheses (Ullman, 1992), decision support tools (Maidin et al., 2019), concept development methods, integrated design activity (Ostergaard & Summers, 2009), design requirements from corporate clients (Gershenson & Stauffer, 1999), design guidance for hypermedia design (Kemp & Buckner, 1999), and comprehensive design (Kemp & Buckner, 1999) in design research (Nyirenda, 2005) have all been classified using taxonomies.

The task at hand was to develop a taxonomy which in a way is bit more holistic. The primary process before we moved onto the development of the taxonomy was to decide the basis of the taxonomy. Sine the problem that we were trying to solve is the lack of a holistic taxonomy, we decided to base the taxonomy on AM features and their utilisation. For better classification we decided to classify the features on the basis of various aspects where the features achieved with AM, enhance the quality of products over conventional manufacturing methods. Eight broad categories were decided, and the various features were classified based on those categories. Due to lack of published materials, the features mentioned have been found using various internet sources. To make the taxonomy more holistic, we have even used materials in the proposed taxonomy. Materials that have been specifically produced for AM or modified for use in AM have also helped immensely in giving AM the added benefits.

### 3.1 Product Classification

The following are the classified products on which the taxonomy was created. Due to a dearth of written materials concerning the items, internet sources were used to research the various initiatives.

## 3.1.1 Acoustics Features

AM has had a significant impact on Acoustics in recent years. Designers have been able to develop newer designs and concepts to improve the quality of sound emanating from speakers. This section discusses the numerous new AM features in an attempt to categorize them.

### *3.1.1.1 Dodecahedron*

Thingiverse user Sean Michael Ragan designed a dodecahedron speaker with twelve speakers, each facing in a different direction as shown in Figure 10. The influence of point-load radiation or sound emitted in all feasible directions can be approximated using this configuration. The speaker in this method allows one to explore the acoustic properties of a room. They are commercially available; however, they are fairly pricey. Some people construct their own, but the complex compound angles and high degree of accuracy and precision required in components make manual tools difficult to work with. It is an excellent application for a 3D printer.

*Figure 10. Dodecahedron Speakers*
**Source:** <*https://www.thingiverse.com/thing:24308*>

### *3.1.1.2 Perpetual Loop*

Almost all conventionally designed and produced speakers exhibit some degree of "back wave" distortion; this occurs when audio impulses fired into the rear of the speaker bounce off the cabinet walls, interfering with signals transmitted to the front of the speaker.

To avoid relying on acoustic absorption, industrial designer Boaz Dekel, a graduate of the Bezalel Academy of Arts and Design, devised an unconventional solution: erase the speaker's back wall (Figure 11). Dekel envisioned a circular speaker in which sound flows "in a continuous self-feeding loop that prevents it from interfering with the signal supplied to the front," according to the Stratasys blog. He was able to rapidly analyse the acoustic response of various geometry and material configurations and identify which were most suitable for speaker cabinets using 3D printing. Other manufacturing or modelling processes, much less within the time constraints required, would not permit such versatility.

The Aleph1's revolutionary self-feeding geometry captures and utilises the back wave's acoustic energy, producing in a clean, open, and natural sound with remarkable detail and separation.

*Figure 11. Aleph1's Concept Design*
**Source:** <*https://www.stratasys.co.in/explore/blog/2016/3d-printed-stereo-speakers*>

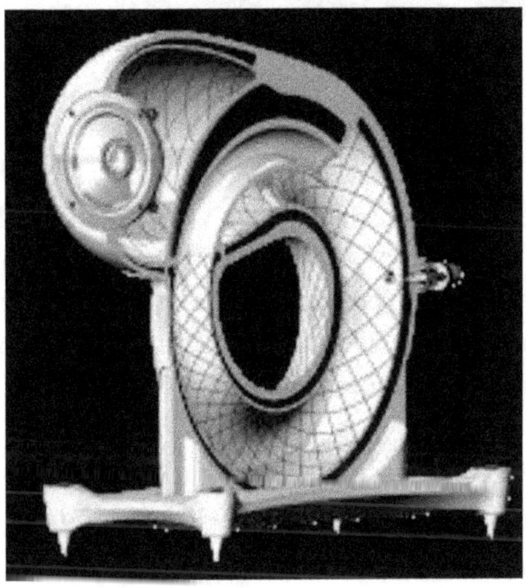

### *3.1.1.3 Anechoic Chamber*

Noise interference must be kept to a minimum in order to enhance the effect of the dynamic sound system. What is the answer? An anechoic sound chamber for the interior of the car, cancelling off external noise and absorbing internal echoes.

The interior trim of a concept car would need to be covered in intricate shapes to create an anechoic chamber, resulting in an irregular surface as shown in Figure 12. The capabilities of that manufacturing technique define the exact intricacy of the shapes that can be made utilising a conventional production method. However, because geometric complexity does not matter in 3D Printing, the designers had previously decided that 3D Printing was the best option for this project. They were, in fact, investigating a specific 3D printing technology: laser sintering. Making that decision early on empowered the designers to construct any designs required for optimal acoustics without regard for complexity-related expenses. Furthermore, being a concept vehicle, this was very certainly a one-time manufacture for which investing in a mould would have been a waste of money.

*Figure 12. PUEGEOT FRACTAL Concept Interior*
**Source:** <*https://www.materialise.com/en/cases/peugeot-fractal-concept-car-3d-printing-acoustic-interiors*>

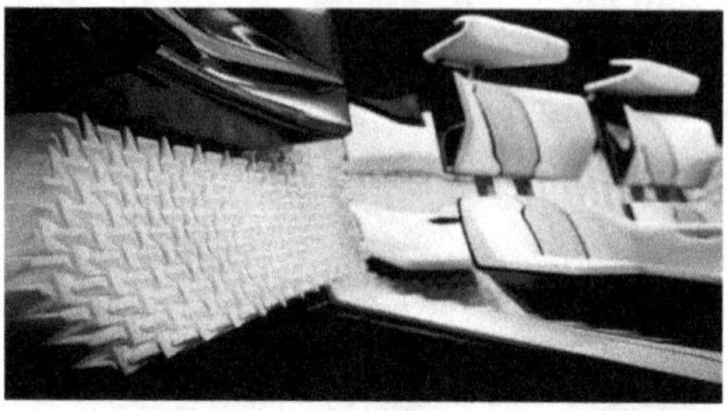

## 3.1.2 Ergonomics Features

The ergonomics of the parts produced have been greatly influenced by additive manufacturing. AM has become the preferred mode of manufacturing in many industries due to its capacity to customize parts based on the needs of the user, as well as its speedier, hassle-free, and automated production of parts. In this section, we will go over the many ergonomic characteristics of AM.

### *3.1.2.1 Personalization*

AM is critical to the ergonomics of the product produced. The product development process, encompassing product planning and production stages, is time-consuming using the traditional manufacturing technique. As a result, personalization with the traditional production method is nearly impossible. AM production is totally automated because it is done directly from the CAD models of the part, resulting in a very rapid product. All these aspects allow the items to be tailored to the individual demands of the consumer, resulting in high levels of customer satisfaction.

### *3.1.2.2 Aesthetics*

The power of AM to create any type of shape, no matter how intricate, is unquestionable. When compared to the traditional manufacturing process, the design freedom associated with AM is far greater, allowing designers to experiment with designs and customise goods to get a more aesthetic feel.

### *3.1.2.3 Improved Product Quality*

When compared to the traditional manufacturing method, the amount of manual labour involved with AM throughout production is substantially lower. This eliminates the possibility of manual errors during production and allows the products to be manufactured with greater accuracy. As a result, we can clearly state that the use of AM improves product quality greatly.

## 3.2. Faster Production

As previously stated, the AM process is virtually fully automated. With the exception of post-processing on the part, no manual labour is required for production. Thus, regardless of how complex the production is, the time required is substantially shorter, and the produced parts may be brought to market extremely rapidly.

### 3.2.1 Safety Features

AM has proven to be a godsend for customers in terms of safety. Companies have been able to develop improved systems and features for increased safety because to increased design flexibility and expertise. In this section, we will go over the many safety aspects of AM.

#### *3.2.1.1 Structural Layering*

Kupol designed the Kollide Safety System, a revolutionary safety system as shown in Figure 13. What is the mechanism at work? This system is made up of three layers of soft, flexible, and lightweight material with an open structure that generates an entire aeration network. As a result, this structure ensures that the

*Figure 13. KUPOL AM helmet*
**Source:** < *https://www.sculpteo.com/blog/2018/10/19/the-world-first-3d-printed-helmet-by-kupol/*>

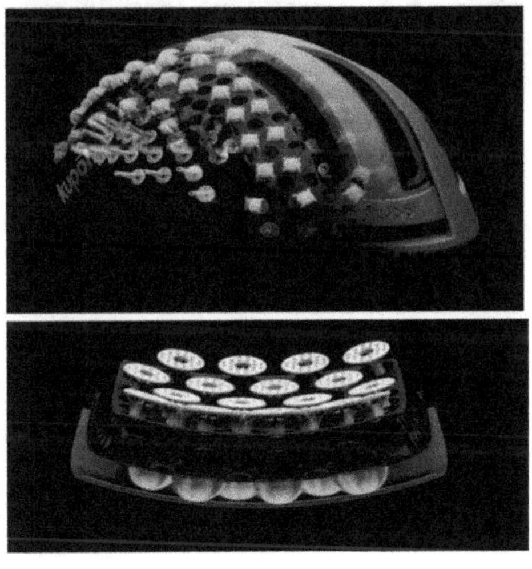

protection is breathable. The Kollide Safety System is a three-layer soft, flexible, and lightweight material system with an open structure that generates a unique aeration network and ensures breathable protection. Using normal approaches, this is not possible. Thus, the use of AM is employed.

### 3.2.1.2 Hexagonal Structures

Under impact force, hexagonal cells outperform EPS foam when surrounded by spherical objects such as human heads. When subjected to high impact, the foam hardens to the point where it prevents an impact and transmits forces through the material, which is not ideal for a helmet. Hexagonal cells as shown in Figure 14., on the other hand, buckle, bend, and distort as a payload decelerates against impact, reducing peak impact forces. Consider it a collection of micro crumple zones for your skull, similar to those found on automobiles. When you hit a HEXR helmet, the contact area grows, resulting in more "cells." When additional cells are recruited, impact forces are reduced, and energy is distributed over a broader area. In reality, this means that forces are reduced at the point of impact rather than spread throughout the medium, as EPS foam does. The standard injection moulding process is capable but accomplishing the complicated design of the helmet would be difficult and certainly not inexpensive. Furthermore, the production process would have to be outsourced. As a result, AM was utilised to compensate for these shortcomings.

*Figure 14. HEXR Am Helmet*
**Source:** *<https://thenextweb.com/news/this-custom-3d-printed-bicycle-helmet-is-the-future-of-saving-skulls>*

### 3.2.1.3 Stress Shielding

Medical device manufacturers can also build implants that resemble a patient's bone rigidity and density. Apart from osseointegration enhancement, 3D-printed implants can significantly minimize stress shielding and enhance physical performance.

Stress shielding is a mechanism in which metal implantation relieve a patient's bone of natural stress. As a consequence, bone density declines, resulting in a weakening of the bone. Stress shielding can result in fractures and dislocations. As a result, creating an implant that is as near to the bone as feasible is crucial for reducing stress shielding and mitigating these adverse effects. Altair, an information technology company, recently combined 3D printing with topology optimization tool to develop a better hip stem implantation. Topology optimization software was utilised to build a novel hip implant design by entering information such as size, weight, and expected implant stress. Compared to generic implants, the improved design distributes strain and stress more efficiently. Additionally, topology optimization is lighter. The improved implant provided 50.7 percent less stress shielding when tested. Simultaneously, its software assisted in determining where to replace the material with lattice structures in order to improve the implant durability limit to approximately 10 million cycles.

## 3.2.2 Comfort Features

Using additive manufacturing for the creation of diverse consumer goods allows us to improve the comfort of the products. This comprises not just design characteristics, but also materials that aid in producing the desired effect.

### *3.2.2.1 Individual Cushioning.*

Individual cushions provide optimum fit, protection against rapid rotational movements, and are a cost-effective alternative to ordinary polystyrene. Individual cushions alter in design to increase comfort and eliminate pressure points.

*Figure 15. Nike Flyprint (AM Upper)*
**Source:** *<https://amfg.ai/2019/09/18/application-spotlight-3d-printing-for-footwear/>*

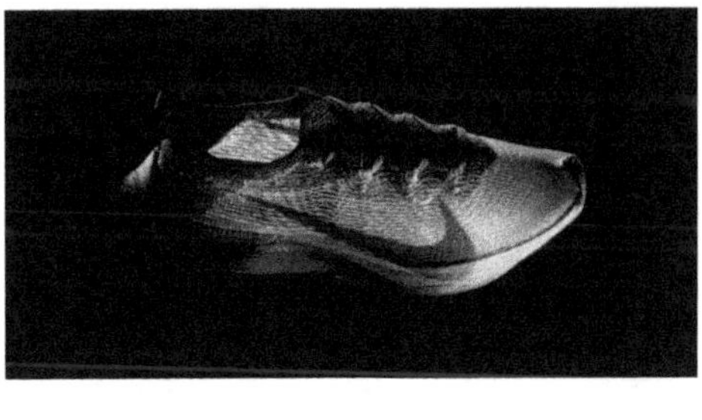

#### *3.2.2.2 Lower Frictional Resistance*

Uppers are typically made of cloth, which can be difficult to manufacture with polymer 3D printers. Some shoemakers, on the other hand, have developed methods for manufacturing uppers out of flexible materials like TPU. Consider the Nike Flyprint as shown in Figure 15, the first sports sneaker with a 3D-printed cloth upper. Flyprint uppers were originally shown last year, and they are manufactured via Solid Deposit Modelling (SDM), which entails melting a TPU filament and building it in microscopic layers. Nike's 3D-printed uppers are more durable than conventionally woven uppers as the layers are fused, the frictional impedance characteristic of knitted and woven textiles is reduced.

### 3.2.3 Enhanced Functionality Features

AM has helped us to develop new concepts to improve the functionality of old goods. This is accomplished by either developing fresh items or making improvements to those that are already in use. In this section, we will go over the numerous aspects that aid in obtaining enhanced functioning.

#### *3.2.3.1 Trifurcating Channel*

GE Research creates a heat exchanger for power generation equipment that has a higher operating temperature and greater thermal efficiency. A trifurcating channel network in a GE heat exchanger as shown in Figure 16., transports hot air from a gas turbine. This network is linked to a second channel network that is filled with colder working fluid and runs in the opposite direction. Although hot air and cool fluid do not interact, their proximity allows for good heat exchange. The researchers concluded that it was only possible using 3d printing. The 3D-printed heat exchanger is made with a special high-temperature, crack-resistant nickel superalloy created by GE Research specifically for this technology. Combining

*Figure 16. Trifurcating channel (Heat Exchanger)*
**Source:** < *https://amfg.ai/2019/09/18/application-spotlight-3d-printing-for-footwear/*>

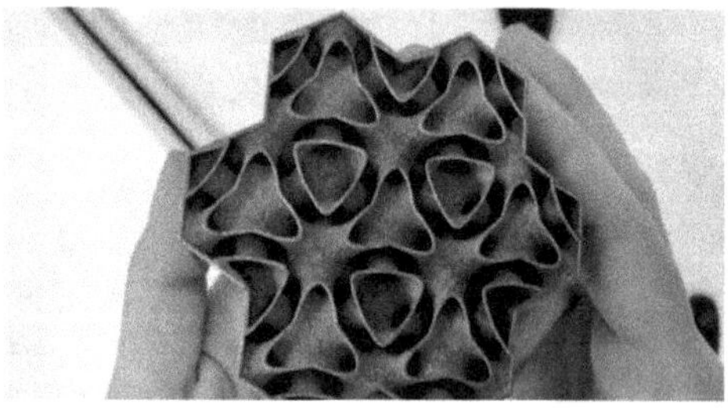

the design freedom of 3D printing with the strength of superalloys is anticipated to result in a significant improvement in heat exchanger performance.

#### 3.2.3.2 Curve flow Channel

The interior geometry of the hydraulic component can be modified using 3D printing to maximise fluid flow and reduce pressure drop. Engineers can instal fluid flow channels in various shapes and sizes, such as manifolds, precisely where they are needed. This means that flow channels can have curved geometries and closer spacing than traditional manifolds, resulting in a more compact and lighter end product. Curved flow channels could increase flow efficiency by 30–70%.

#### 3.2.3.3 Roller Cage

Bowman constructs the Rollertrain cage with HP's Multi Jet Fusion technology and nylon material (PA11). The cage has an interlocking structure that pins each cage segment together using rolling components. This design allows the cage to hold up to 45 percent more rollers than standard versions. The greater the number of rollers, the more weight can be distributed among more rolling sections. This results in a 30-40% increase in load capacity and a threefold increase in cage life. Rollertrain cages are the same price as typical parts. However, significantly improved performance and longevity add value, making 3D-printed bearings preferable to traditional designs.

#### 3.2.3.4 Trabecular Structure

The human body contains two types of bone tissue: trabecular bone and osseous bone. It has a squishy, permeable texture that is unmatched by an implant. Historically, implant producers have created trabecular implant structures using a special coating. However, this increases the risk of structural failure of the implant. On the other hand, 3D printing can rapidly fabricate implantation with a trabecular structure, obviating the need for a coating phase. More crucially, when combined with a 3D printing, the trabecular, porous structure results in a eventual stronger implant with a lesser risk of structural failure. Such design freedom is enabled by metal 3D printing methods such as Selective Laser Melting and Electron Beam Melting. These methods employ a high source of energy such as a laser (SLM) or an electron beam to deposit a thin layer of compatible titanium powder particles (EBM). Performing this method numerous times can result in a complex implant that responds substantially better to the human skeletal structure than traditionally manufactured implantation.

### 3.2.4 Lightweight Features

We can lower the weight of the produced items by using AM into the manufacturing process. Reducing the weight of the items greatly improves the performance of the parts. This section discusses the many elements that aid in the promotion of the assets.

#### *3.2.4.1 Honeycomb*

Honeycomb structures are one of the most common structures used to reduce product weight. The standard manufacturing technique is hard and time-consuming for producing Honeycomb structures, whereas AM solves the problem rapidly.

#### *3.2.4.2 Hollow Structures*

We can optimise the general topology of the manufactured part using AM. According to research, parts made with hollow structures and optimised topology perform better than entirely solid geometries. Hollow structures allow designers to minimise the weight and make the shape lighter. Raw materials for those specific regions are also saved. Furthermore, this contributes to lowering the overall cost of the production process.

#### *3.2.4.3 Scalmalloy*

Scalmalloy offers extremely unique properties for metal additive manufacturing. Its material qualities combine the low weight of aluminium (AlSi10Mg) with about equal specific load bearing capacity and dcutility (Ti6Al4V). Scalmalloy exhibits a higher level of impedance to corrosion and a more robust structure at elevated temperatures than other AM aluminium alloys, making it suitable for a wide variety of high-performance applications. It was developed primarily for additive manufacturing, more precisely for Selective Laser Sintering (SLS) or Direct Metal Laser Sintering (DMLS).

#### *3.2.4.4 MMC*

Not every metal lends itself to the additive process. As it is built up in a laser powder-bed additive machine, a metal shape breaks apart due to internal strains. The effect is different when ceramic is applied to metal—a ceramic powder that does not melt during construction. This metal can be 3D-printed thanks to the ceramic and the company's proprietary procedure for adding this material. More importantly, the combination of ceramic and metal creates a strong, lightweight, heat-resistant material with a wide range of future applications, including the additive fabrication of significantly lighter aerospace or vehicle engine components.

## 3.2.5 Flexibility Features

The use of additive manufacturing in part production can improve the flexibility of the parts produced. This section discusses the various features that contribute to the increased flexibility of the parts produced.

### *3.2.5.1 Auxetics*

Auxetics refer to those special kind of lattice structures that have a negative poisons ratio. Having a negative poisons ratio, the lattice structure gets enhanced properties in the form of increased flexibility and shock absorbing capabilities. With the type of construction required to produce the material, it is only possible to be made with the help of AM.

### *3.2.5.2 Variable Density Lattice*

AM enables footwear manufacturers to experiment with and incorporate new design elements. Consider the midsole: it is frequently made as a single solid component that provides the same level of support throughout the shoe. Because AM can create midsoles with lattice structures, as shown in Figure 17, that would be difficult to injection mould, shoe performance can be significantly improved. We can specifically engineer these materials to introduce varying densities within the midsole. Designers can improve cushioning across the shoe by adjusting different midsole components, resulting in better-performing footwear.

*Figure 17. 3D printed Midsoles.*
**Source:** < *https://amfg.ai/2019/09/18/application-spotlight-3d-printing-for-footwear/*>

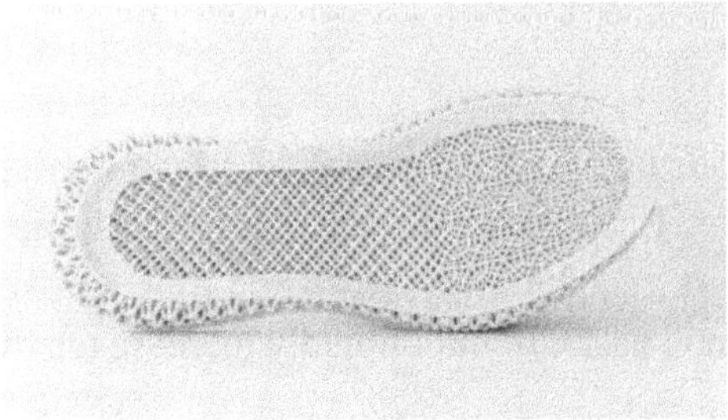

#### *3.2.5.3 Thermoplastic Polyurethane*

Thermoplastic polyurethane (TPU) is an elastomer with high processing flexibility and durability, combining the properties of both thermoplastics and rubbers. We discover in its chemical composition that its adaptability is related to alternate sequences of hard and soft segments, i.e., changing the percentage of these segments changes the material's hardness and flexibility. This has an impact on the final part's transparency, touch softness, and adherence. Overall, we can say that TPU is a very diverse polymer that provides the parts with an intriguing set of properties. This also opens up the possibility of 3D printing flexible models. What, on the other hand, should we consider when employing TPU? In the additive manufacturing sector, this material opens up a world of possibilities for creating tyres and shock absorbers for a variety of applications such as footwear, elastic soles, and the automotive industry. TPU is ideal for end-of-life components, prototypes, concept models, and custom components. This type of material, for example, is widely used to make mobile phone covers that protect the device from shocks and fractures.

#### *3.2.5.4 Silicone*

It has exceptional elastomer properties, with a wide range of hardness variation, making it ideal for applications requiring moderate compression or high flexibility. The material is also biocompatible, making it an excellent choice for medical applications. Previously, the only way to produce silicone parts was through traditional manufacturing, i.e. injection moulding, which is very expensive in its true sense. Production on a large scale is possible with the traditional injection moulding process, but customization is not. AM has solved the problem. Am allows for the extensive use of Silicone, which aids in imparting the required flexibility to the produced product.

### 3.2.6 Consolidation Features

AM allows products to be manufactured in a single step, reducing the need for assembly. This section discusses AM's various consolidation features.

#### *3.2.6.1 Brackets*

Some brackets are made up of several pieces, which may lengthen the time it takes to assemble the bracket. Because of AM, the bracket can be designed and produced as a single unified part, requiring less work and time to assemble. Furthermore, AM a bracket as a single part will undoubtedly increase the overall strength of the

*Figure 18. Evolution of the Brackets.*
**Source:** *<https://www.materialise.com/en/cases/philips-lightbulb-moment-3d-printing-becomes-essential-production-thinking>*

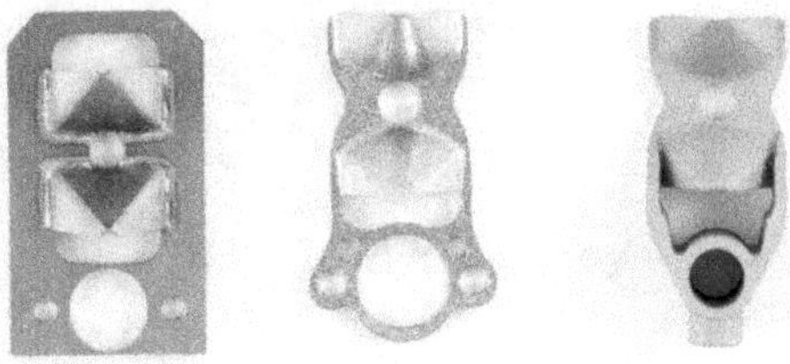

part. Figure 18 shows how the brackets have evolved over the years, from being produced in parts to being produced in a single unit with the application of AM.

### *3.2.6.2 Frames*

Steel has long been the preferred bicycle material. However, in recent years, lightweight materials such as titanium and carbon fibre have emerged, which, among other advantages, can help reduce the overall weight of a bike frame. A lightweight bike offers numerous advantages to cyclists. For starters, it enables riders to ride at higher speeds and with greater ease. Second, the lighter a bike is, the easier it is to transport. Finally, when riding, a lighter bike may be easier to manage and respond to a rider's actions, giving competitive cyclists significant advantages. Creating titanium or carbon fibre motorcycles using traditional manufacturing methods is difficult, not least because of the time-consuming and often labour-intensive manufacturing processes. The most significant of these challenges is that the production of these light-weight frames is traditionally done in batches, which makes the process extremely labour-intensive. We have the luxury of producing these in one go with AM, eliminating the need to join the various components together.

### *3.2.6.3 Interconnects*

Interconnects are electronic structures that connect two or more circuit elements (such as transistors). Current interconnect manufacturing methods, such as wire bonding, have drawbacks such as long conductor routes and high mechanical stress on sensitive components. Printing interconnections directly on PCB and RF component pads as shown in Figure 19, may be able to solve these problems. Optomec's Aerosol Jet technology is one of the technologies that can print conformal interconnections on 3D surfaces, eliminating the need for wire bonding.

*Figure 19. AM Interconnects.*
**Source:** *<https://www.3dprintingmedia.network/northrop-grumman-study-performance-reliability-3d-printed-electronics/>*

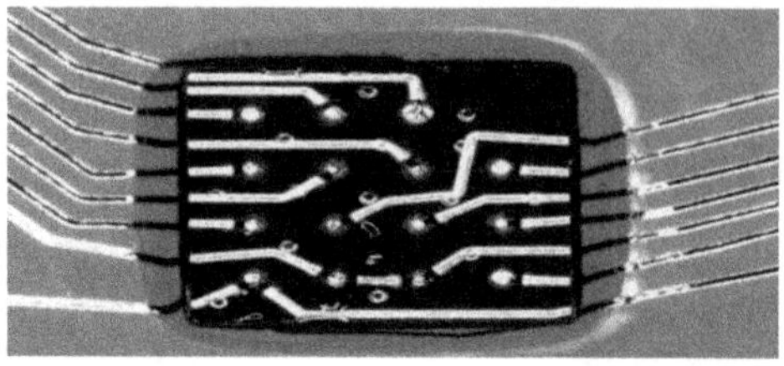

The features of the taxonomies have been discussed in detail in this section. We attempted to categorize the features of AM in general using a classification based on the advantages that AM has over traditional manufacturing processes. It gives us a general idea of the characteristics that give AM an advantage over traditional manufacturing processes.

## 3.3 Developed Taxonomy

The features classified with the help of the products have all been a direct result of the advantages of AM process. With the ease in the manufacturing process and freedom with the design process, the features have been achieved from the products classified, we have come up with the taxonomy for AM production as shown in Figure 20.

*Figure 20. Feature Taxonomy for AM*

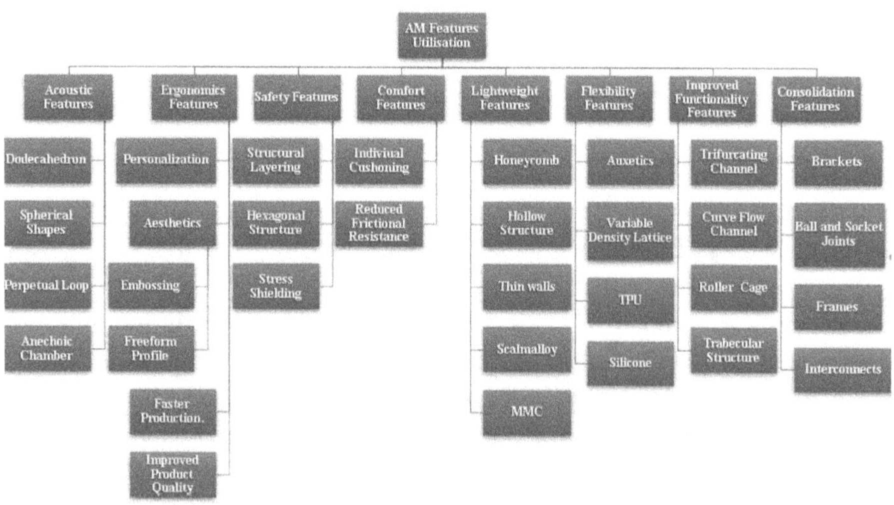

The taxonomy developed is different from the ones already present. The taxonomy classifies AM utilization features based on what advantages AM brings over the conventional manufacturing process and what features are responsible for the discussed advantage. The taxonomy is not only limited to shape features but covers a variety of features, including products and materials as well.

With the taxonomy developed, we look forward to the development of an algorithm for the part placement in the AM enclosure. The two aspects i.e., the development of the taxonomy, and the placement of parts are very important individually. The strategies present already cannot be used directly for the placement of parts. Hence, the section concerns itself with the development of a custom algorithm for part placement.

## 4. PART PLACEMENT

Features are the aspects that enable the manufacture of parts using AM, but it is the part placement that optimizes and makes the entire process feasible. Therefor it is an equally important part of the entire process. The part placement or nesting as it is sometimes referred to is an age-old practice that has been applied in various domains like ship building, textiles, etc. The age old process of nesting can also be put into good use in various other fields such as additive manufacturing, and 3D printing. Nesting can be used to improve tool movement and in maximizing how many pieces can be fabricated into a single build session. The main difference that we face here is that 3D parts often have a cross section that changes with height, which can cause interference between adjacent parts as they are built up.

Part placement in AM is extremely important which focuses mainly on optimizing a single part so as to improve the part quality, minimize the time taken and total cost. However, due to the continuous innovations that are incorporated and the expanding market. AM has now grown to a stage where it is able to provide small batches of customized. With different types of AM process, the challenges of part placement are different. For processes such as SLA, EBM etc., the requirement of support structures is negligible. On the other hand, with processes such as SLS, the placement of parts requires support structures and thus can only be placed in one layer on the build volume.

Before moving on to the actual problem itself it is necessary to understand into what all contexts the problem can be classified. The nesting problem can be classified on the basis of placement into 3 main groups into full, subset, and uncertain placement context.

a) **Full placement context:** Small batches of parts are produced in turns since requests for manufacturing administration might be received in small batches consecutively.
b) **Subset placement context:** A significant quantity of parts is waiting to be made by a machine, but the build volume can only accommodate a subset of the part group. There are two major scenarios in real-world production that fit this description. One is how to nest or cram subset pieces into the current machine run as much as feasible, and the other is how to reduce the total number of build runs required to make the entire part group.
c) **Uncertain placement context:** It is unknown if a group of pieces can be filled into a particular build volume or not, even if the group's total volume is less than the build volume.

Accordingly, these can be categorized into 4 parts to suite the 3 groups i.e. Modified strip packing, Knapsack, Bin packing, and Decision problems.

a) **Modified 'Strip Packing' problem:** The 'Strip Packing' problem is defined as follows: given a set of shapes and a width W, find the shortest length of a rectangular region that contains all shapes. For part placement in the context of full placement, a modified 'Strip Packing' issue can be employed. The description can be as follows: given a set of parts and a build volume, maximise the beneficial value or values indicating the user or production preference while fitting all the parts into the specified build volume.
b) **Knapsack problem:** The classic Bin Packing problem is as follows: given a collection of forms and a collection of regions, reduce the number of regions required to place all shapes. As a result, this problem can be used to illustrate the second situation in the context of subset placement, as all components should be built. However, the purpose of this challenge may change as real-world application requirements in AM vary. Generally, the purpose is to limit the number of production runs for an additive manufacturing machine.
c) **Bin pack problem:** The classic Bin Packing problem is as follows: given a collection of forms and a collection of regions, reduce the number of regions required to place all shapes. As a result, this problem can be used to illustrate the second situation in the context of subset placement, as all components should be built. However, the purpose of this challenge may change as real-world application requirements in AM vary. Generally, the purpose is to limit the number of production runs for an additive manufacturing machine.
d) **Decision problem:** The objective of this challenge is to determine whether a collection of shapes fits into a certain region. It is appropriate to define the part placement problem in an unclear context of part placement. Due to the

complicated geometry of 3D models, it is extremely difficult for operators or technicians to determine whether a set of pieces with a total volume less than or equal to a particular build volume can fit within the build volume or not, much alone discover an effective nesting or packing solution.

## 5. ALGORITHMS FOR PART PLACEMENT

Algorithms refer to the sequential steps and processes that should be followed to solve a particular problem. They are one of the tools that can be used to aid in problem-solving. There are various kinds of algorithms devised to solve various problems. Algorithms that use a similar problem-solving approach can be grouped. Some of the various types of algorithms are Genetic Algorithms, Dynamic Programming Algorithms, Greedy Algorithms, Dragonfly Algorithm, etc.

### 5.1 Genetic Algorithm

Genetic Algorithms (GA) are search-based algorithms that incorporate notions from natural selection and genetics. The term "GA" refers to a subset of a much broader field of computation known as Evolutionary Computation. GA was invented by John Holland, and it has subsequently been successfully used to a variety of optimization problems by his students and colleagues at the University of Michigan, most notably David E. Goldberg. In GA, we have a collection of probable solutions to a particular problem, or a population of alternative answers. These solutions are then subjected to recombination and mutation (as in natural genetics), resulting in the birth of new children. Each person (or candidate solution) is assigned a fitness value (depending on the value of its objective function), and the fitter individuals are given a higher probability of mating and producing more "fit" individuals. This is consistent with Darwin's "Survival of the Fittest" theory. Thus, we continue "evolving" better individuals or solutions through generations until we reach a threshold for halting.

### 5.2 Dynamic Programming Algorithm

Dynamic programming is a technique for mathematical optimization as well as computer programming. Richard Bellman created the approach in the 1950s and it has found applications in a wide variety of industries, from aeronautical engineering to economics. In both cases, it refers to the process of recursively simplifying a big problem by breaking it down into simpler sub-problems. While some choice issues cannot be disassembled in this manner, decisions spanning multiple points in time frequently do so recursively. Similarly, in computer science, an issue is said to have an optimal substructure if it can be solved optimally by decomposing it into sub-problems and then recursively discovering the best solutions to the sub-problems. While this method is helpful in that it always produces the correct answer, it becomes extremely lengthy and computationally intensive as it evaluates and calculates all conceivable outcomes.

## 5.3 Greedy Algorithm

Greedy Algorithm is a problem solving approach with an algorithmic strategy that makes the best optimal decision from the available choices at each stage, with the hopes that this will eventually lead to a globally optimum solution. This means that the Algorithm picks the best solution at the moment without regard for consequences. It picks the best immediate output but does not consider the big picture. Hence it is considered greedy. This method is advantageous due to its sheer simplicity and easiness. It also guarantees the optimal solution in problems such as the minimum no of coins to give while making change.

## 5.4 Dragonfly Algorithm

Dragonfly algorithm (DA) is a novel swarm intelligence meta-heuristic optimization algorithm inspired by the dynamic and static swarming behaviours of artificial dragonflies in nature. It has proved its effectiveness and superiority compared to several well-known meta-heuristics. The main inspiration of the DA algorithm originates from nature itself, copying the static and dynamic swarming behaviors' of dragonflies. The algorithm consists of the following phases namely optimization, exploration, and exploitation. The algorithm is designed by demonstrating the social interaction of dragonflies in navigating, and searching for food (which can be used to find the desirable answer), while at the same time and avoiding enemies (which can be used as the conditions or parameters) when swarming dynamically or statistically. This is a relatively new and complicated method.

None of these algorithms alone can be used directly to solve the problem at hand. However, their approaches can be used in tackling the problem, so we come up with a custom algorithm in order to solve the problem. However, we can use the various approaches that each Algorithm uses to exploit the available data.

## 6. CONCLUSION AND FUTURE SCOPE

AM has been widely accepted as a revolutionary force in manufacturing. As 3D printers become more and more affordable, they will inevitably be used for local, small-scale manufacturing, which will eliminate the supply chain and the high logistical requirements behind many products. AM, due to its layer-by-layer process has opened up new possibilities and has helped in identifying features that were once unnoticed or non-existent. However, most of the classifications present are based on shapes and raw material used, we through this document wish to highlight those classifications can not only be based on shapes and raw material but also on the basis of other aspects such as functionality and safety. The taxonomy developed helps a fair bit in the classification of AM features and gives us an idea about the advantages of AM. The part placement enables the feasible production and overall optimization of the process. Various algorithms can be employed to help in nesting the various. This work can be further improved by combining the features developed with the algorithm so as to orient the parts accordingly.

## REFERENCES

AMFG. (2019). *Application Spotlight: 3D Printing for Footwear*. AMFG. Retrieved from https://amfg.ai/2019/09/18/application-spotlight-3d-printing-for-footwear/

Bandyopadhyay, A., Bose, S., & Das, S. (2015). 3D printing of biomaterials. *MRS Bulletin*, *40*(2), 108–115. doi:10.1557/mrs.2015.3

Bártolo, P. J. (2011). *Stereolithography: materials, processes and applications*. Springer Science & Business Media. doi:10.1007/978-0-387-92904-0

Beedham, M. (2019). *This custom 3D-printed bicycle helmet is the future of saving skulls*. TNW. Retrieved from https://thenextweb.com/news/this-custom-3d-printed-bicycle-helmet-is-the-future-of-saving-skulls

Berman, B. (2012). 3-D printing: The new industrial revolution. *Business Horizons*, *55*(2), 155–162. doi:10.1016/j.bushor.2011.11.003

Boissonneault, T. (2019). *Northrop Grumman study showcases performance and reliability of 3D printed electronics*. 3D Printing Media Network. Retrieved from https://www.3dprintingmedia.network/northrop-grumman-study-performance-reliability-3d-printed-electronics/

Cheng & Huang. (2020). *Preparation and characterization of color photocurable resins for full-color material jetting additive manufacturing*. Academic Press.

Chua, C. K., Leong, K. F., & Lim, C. S. (2010). *Rapid prototyping: principles and applications (with companion CD-ROM)*. World Scientific Publishing Company. doi:10.1142/6665

Cunningham, J., & Dixon, J. (1988). *Designing with features: the origin of features*. In *Proceedings of the 1988 ASME International Computers in Engineering Conference and Exhibition*. ASME.

Derr, R. L. (1973). *A taxonomy of social purposes of public schools: A handbook*. McKay.

Fontana, M., Giannini, F., & Meirana, M. (1999). A free form feature taxonomy'. In: P. Brunet, R. Scopigno (Eds.), *Proceedings of Eurographics 1999, Computer Graphics Forum, Vol. 18*, pp. 107-118.

Gaget, L. (2018). *The world's first 3D printed bike helmet by Kupol!* Sculpteo. Retrieved from https://www.sculpteo.com/blog/2018/10/19/the-world-first-3d-printed-helmet-by-kupol

Gershenson & Stauffer. (1999). *A taxonomy for design requirements from corporate customers*. Academic Press.

Gupta & Gurumoorthy. (2008). *Extending feature based semantic interoperability for sheet-metal product development*. Academic Press.

Gupta & Gurumoorthy. (2013). *Classification, representation, and automatic extraction of deformation features in sheet metal parts*. Academic Press.

Gupta & Gurumoorthy. (2021). *Feature-based ontological framework for semantic interoperability in product development*. Academic Press.

Hull, C. W. (1984). *Apparatus for production of three-dimensional objects by stereolithography.* United States Patent, Appl., No. 638905.

Johnson, D. W., Sherborne, C., Didsbury, M. P., Pateman, C., Cameron, N. R., & Claeyssens, F. (2013). Macrostructuring of emulsion-templated porous polymers by 3D laser patterning. *Advanced Materials, 25*(23), 3178–3181. doi:10.1002/adma.201300552 PMID:23606519

Joshi, Sheikh, & Prototyping. (2015). *3D printing in aerospace and its long-term sustainability.* Academic Press.

Kemp & Buckner. (1999). *A taxonomy of design guidance for hypermedia design*. Academic Press.

Kruth, J.-P. (2003). *Lasers and materials in selective laser sintering*. Academic Press.

Kruth, J.-P. (2007). *Consolidation phenomena in laser and powder-bed based layered manufacturing*. Academic Press.

Levy, Schindel, & Kruth. (2003). *Rapid manufacturing and rapid tooling with layer manufacturing (LM) technologies, state of the art and future perspectives.* Academic Press.

Lim, C. W. J., Le, K. Q., Lu, Q., & Wong, C. H. (2016). An overview of 3-D printing in manufacturing, aerospace, and automotive industries. *IEEE Potentials, 35*(4), 18–22. doi:10.1109/MPOT.2016.2540098

Maidin, Pei, & Campbell. (2019). *A Taxonomy of Design Features for Additive Manufacture*. Academic Press.

Mandrycky, C., Wang, Z., Kim, K., & Kim, D.-H. (2016). 3D bioprinting for engineering complex tissues. *Biotechnology Advances, 34*(4), 422–434. doi:10.1016/j.biotechadv.2015.12.011 PMID:26724184

Materialise. (n.d.a). *The Peugeot Fractal Concept Car: 3D Printing Acoustic Interiors*. Materialise. Retrieved from https://www.materialise.com/en/cases/peugeot-fractal-concept-car-3d-printing-acoustic-interiors

Materialise. (n.d.b). *Philips' Lightbulb Moment: 3D Printing Becomes Essential Production Thinking*. Materialise. Retrieved from https://www.materialise.com/en/cases/philips-lightbulb-moment-3d-printing-becomes-essential-production-thinking

Melchels, F. P., Feijen, J., & Grijpma, D. W. (2010). A review on stereolithography and its applications in biomedical engineering. *Biomaterials*, *31*(24), 6121–6130. doi:10.1016/j.biomaterials.2010.04.050 PMID:20478613

Morris & Stauffer. (1994). *A design taxonomy for eliciting customer requirements*. Academic Press.

Murphy, S. V., & Atala, A. (2014). 3D bioprinting of tissues and organs. *Nature Biotechnology*, *32*(8), 773–785. doi:10.1038/nbt.2958 PMID:25093879

Nalluri, S. (1994). *Form feature generating model for feature technology*. Indian Institute of Science, Department of Mechanical Engineering Bangalore.

Ngo, T. D., Kashani, A., Imbalzano, G., Nguyen, K. T. Q., & Hui, D. (2018). Additive manufacturing (3D printing): A review of materials, methods, applications and challenges. *Composites. Part B, Engineering*, *143*, 172–196. doi:10.1016/j.compositesb.2018.02.012

Nyirenda & Bronsvoort. (2009). *A framework for extendable freeform surface feature modelling*. Academic Press.

Nyirenda, P.J. (2005). *A generic taxonomy for defining freeform feature classes*. Academic Press.

Ostergaard & Summers. (2009). *Development of a systematic classification and taxonomy of collaborative design activities*. Academic Press.

Pearson, A. (2020). *How Stratasys 3D printing completely turned around stereo speaker acoustics*. Stratasys. Retrieved from https://www.stratasys.co.in/explore/blog/2016/3d-printed-stereo-speakers

Pernot, J.-P. (2008). *Incorporating free-form features in aesthetic and engineering product design: State-of-the-art report*. Academic Press.

Pham, D. T., & Gault, R. S. (1998). A comparison of rapid prototyping technologies. *International Journal of Machine Tools & Manufacture*, *38*(10-11), 1257–1287. doi:10.1016/S0890-6955(97)00137-5

Ragan. (2012). *Makerbot Thingverse*. Retrieved from https://www.thingiverse.com/thing:24308

Raman, R., & Bashir, R. (2015). Stereolithographic 3D bioprinting for biomedical applications. In Essentials of 3D Biofabrication and Translation. Elsevier. doi:10.1016/B978-0-12-800972-7.00006-2

Rochus, P. (2007). *New applications of rapid prototyping and rapid manufacturing (RP/RM) technologies for space instrumentation*. Academic Press.

Schubert, Van Langeveld, & Donoso. (2014). *Innovations in 3D printing: a 3D overview from optics to organs*. Academic Press.

Shneiderman & Norman. (1992). *Questionnaire for User Interface Satisfaction (QUIS), Designing the User Interface: Strategies for Effective Human-Computer Interaction*. Academic Press.

Sinha, R. P., & Häder, D.-P. (2002). UV-induced DNA damage and repair: A review. *Photochemical & Photobiological Sciences*, *1*(4), 225–236. doi:10.1039/b201230h PMID:12661961

Spangenberg, A. (2013). Recent advances in two-photon stereolithography. *Updates in Advanced Lithography*, 35-63.

Ullman. (1992). *A taxonomy for mechanical design*. Academic Press.

Weiß, T. (2009). Two-photon polymerization for microfabrication of three-dimensional scaffolds for tissue engineering application. *Engineering in Life Sciences*, *9*(5), 384–390. doi:10.1002/elsc.200900002

Wilson, P., & Pratt, P. (1988). A taxonomy of features for solid modeling. Geometric modeling for CAD applications. Elsevier Science Publishers.

Wohlers, T. T. (2009). *Wohlers report 2009: state of the industry: annual worldwide progress report*. Wohlers Associates.

Yadroitsev, I. (2007). *Strategy of manufacturing components with designed internal structure by selective laser melting of metallic powder*. Academic Press.

Yap, Y., & Yeong, W. (2014). Additive manufacture of fashion and jewellery products: a mini review: This paper provides an insight into the future of 3D printing industries for fashion and jewellery products. *Virtual and Physical Prototyping*, *9*(3), 195–201. doi:10.1080/17452759.2014.938993

Zhang, Y. (2016). *Feature based building orientation optimization for additive manufacturing*. Academic Press.

Ziemian & Crawn. (2001). *Computer aided decision support for fused deposition modeling*. Academic Press.

# Chapter 8
# Process Optimizations of Direct Metal Laser Melting Using Digital Twin

**Sachin Salunkhe**
*Vel Tech Rangarajan Dr. Sagunthala R&D Institute of Science and Technology, India*

**Vishal Gangadhar Naranje**
https://orcid.org/0000-0003-1076-6803
*Amity University, Dubai, UAE*

**Jayavelu S.**
*Vel Tech Rangarajan Dr. Sagunthala R&D Institute of Science and Technology, India*

**Atiq Rehman**
*Amity University, Dubai, UAE*

## ABSTRACT

*Over the past decades, air traffic has increased to such an extent that it has highly impacted (anthropogenic) climate change due to heat, noise, particulates, and gas emissions. With airplane turbines being a pivotal contributor to such adverse developments, there has been an increasing interest in research regarding the optimization of airplane turbines. In line with these efforts, this chapter adopts an innovative approach that bridges the digital and physical through the application of digital twin technology to direct metal laser melting to optimize product development. Specifically, it encompasses a guideline towards how digital twin solutions are created based on all the latest research. A manual approach devises a digital twin interface where the prototype is manufactured used additive manufacturing. This manual can then be applied to optimize airplane turbines regarding their safety, environmental impact, fuel efficiency, and cost.*

DOI: 10.4018/978-1-7998-8516-0.ch008

## INTRODUCTION

While there are many products in the manufacturing process where quality is of critical importance, there is hardly an industry where this applies more than the aerospace industry, where the stakes of defective outputs are extremely high so that even minute production errors can decide over life and death. In 2005, Chalk's Ocean Airways Flight 101 caused 20 people to lose their lives when the plane crashed in Miami, USA, due to an in-flight wing breakage (Aviation Safety Network, 2005). The environmental impact of air traffic has been shown to be a major driver of climate change, not only because of its CO2 emissions, but also because of heat, noise, particulate and gas emissions that interact with the atmosphere by aircraft engines, which together account for a whopping 3.5% of all anthropogenic climate change, according to recent estimates by the Intergovernmental Panel on Climate Change, IPCC (European Union Aviation Safety Agency, 2019). As a result, it becomes even more critical to innovate in the field of aeroplane design, both in terms of sustainability and fuel efficiency, as well as the quality, safety, and durability of all individual components within an aeroplane.

One critical way to achieve such quality improvements is to have a prototype created using additive manufacturing, more commonly referred to as 3D printing. Rapid prototyping is useful in the aviation industry because it allows designers to quickly visualise, test, and adjust product concepts, all of which are critical to the final product's success. Even beyond rapid prototyping, additive manufacturing is an important tool in the production process because it bypasses the otherwise time-consuming and expensive process of experimenting with different combinations of process variables such as the heat source, power distribution, hatch spacing, or substrate preheats (DebRoy, Zhang, Turner, & Babu, 2017). Overall, additive manufacturing is a revolutionary production process as it a) drastically reduces the cost and time used to optimize otherwise expensive trial and error process in product design; b) enhances the baseline conditions for product qualification; and c) reduces the wastage of material and increases overall quality of the product because the number of defects decrease as a result of additive manufacturing.

### Enhancing Additive Manufacturing through Digital Twin

Additive manufacturing has 3 key disadvantages that currently hold back its potential. First, new product designs from rapid prototyping are often met with distortions in the printing process. In other words, the technology is not yet developed thoroughly to an extent where 3D-printed parts are always accurate in the first try. Second, the quality of workpiece is still often substandard, which is an important consideration in high stake industries such as that of aerospace. Finally, as a result

of these design outcomes, substantial amounts of time and feedstock material is wasted in the process that 3D printing was initially set out to save. Because of the increasing demands, the aerospace industry must innovate in terms of sustainability and durability of various parts, as illustrated by this brief overview. This enormous challenge aims to combine a third technology that addresses both issues at once: a digital twin. Virtual models of processes, products, and services are known as digital twins. Digital twins can therefore bridge the physical and virtual world in additive manufacturing to analyse and monitor all the data required to make the above improvements in aerospace. When an additive manufacturing digital twin is available, any of the problems listed above can be addressed in advance, allowing for new opportunities to be explored without wasting valuable resources. In this book chapter, we will focus specifically on the development of an airplane turbine for two reasons. First, from a safety perspective the turbines are the key part of an airplane that accounts for its safe takeoff, flight, and landing in the context of a changing (and challenging) range of different temperatures, altitudes, gross weights, and payload conditions (Boeing, 2019). A second reason is that the aeroplane turbine is at the centre of environmental improvements. Turbine optimizations, according to leading aeroplane engineers, could reduce engine noise by 50-75%, resulting in a remarkable 12:1 bypass ratio, and increase fuel efficiency by up to 25% in the next five years (Aviation Pros, 2012; Boeing, 2019). Creating a digital twin of additive manufacturing (as shown in Figure 1), if used in the aerospace industry, would be an ideal solution to the current conundrum in aeroplane product development because it "holistically integrates models for temperature, microstructure and properties, and residual stresses and distortion that further consider defects" (DebRoy, Zhang, Turner, & Babu, 2017, p. 123).

## Current State of Digital Twin Technology

Given the rapid opportunities that digital twin technology enables across all industries, it is widely regarded as a revolutionary process (Lim, Zheng, Chen, 2019). Significant progress has been made in this area in recent years, indicating that the completion of a fully functional first-generation digital twin of additive manufacturing is possible (Lua, Liub, Wangc, Huanga, & Xua, 2020). The scientific community anticipates a rapid increase in physically accurate digital twin models in the near future as a result of a combination of three critical factors (Sun, Bao, Li, Zhang, Liu, & Zhou, 2020). To begin, computational efficiency has increased exponentially as a result of ground-breaking advances in solvers, preconditioners, and adaptive methods. Second, this overall increase in computational efficiency has been absorbed by additive manufacturing machines. Thirdly, the underlying infrastructure serves as the ultimate catalyst for the aforementioned changes. According to DebRoy, Zhang,

*Figure 1. Diagram of innovation.*

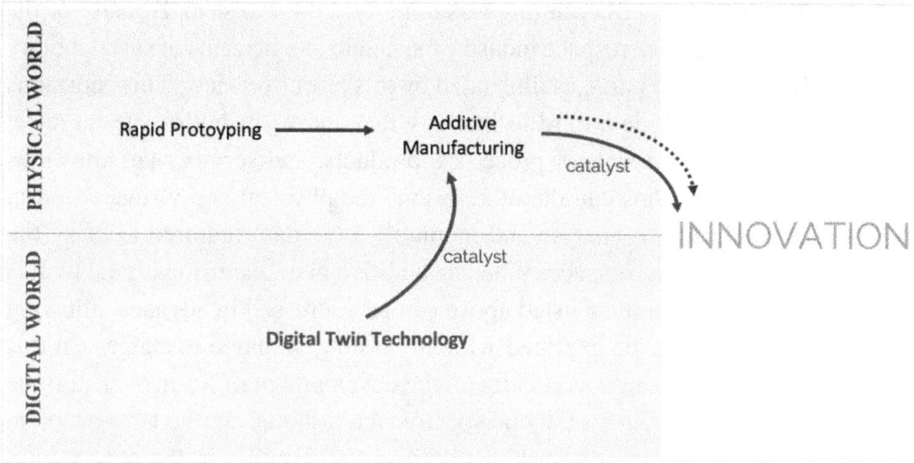


Turner, and Babu, "today's computationally efficient, high-fidelity models can simulate the most critical factors affecting the properties of additive manufacturing products and, once validated, can serve as digital twins for 3D printing machines" (2017, p. 119). The book chapter emphasises the importance of additional research into this emerging technology in order to combat climate change and contribute to exponentially positive effects across industries for product design.

## METHODOLOGY: DESIGN AND IMPLEMENTATION

Digital twin of an additive manufacturing tool used to airplane turbines need to be designed, several steps need to be taken. First, data needs to be accumulated and placed in structured data sets, such as microstructure properties or thermal stresses, which are specific to the final analysis of the product. All of these data sets will then be used to create boundary conditions on which the product designs and simulation build. As such, the second step is the design of the product, which in this case is an airplane turbine, using CAD/CAM where the desired product geometry is set and all the gathered data is allotted to the specifics of that geometry. Using any CAD software, a virtual representation of the final product is created. Based on all the acquired semi-analytical solutions, the software is fed with different calculated boundary conditions after which the product undergoes several simulations to test for optimization. After these steps are taken, all of the data acquired that includes geometric magnitudes, boundary conditions, simulation analysis, and optimized results are accumulated into a single operating system to form the first interface

of a digital twin. This interface is then bridged to CAM or additive manufacturing machines, which then continue to print and therefore manufacture the desired product for optimized results. After the first prototype is manufactured, it undergoes stringent physical analysis whereby the data derived from the analysis is then compared to the optimized virtual data that has previously been established. At this point both data sets are merged and the system is taught to adapt and learn any errors or deviances that have occurred using artificial intelligence integrated systems. Once this system is set, it is synched with the initial digital twin interface, which completes the finalization of a digital twin for additive manufacturing. A detailed insight into these steps is outlined in the subsequent sections.

## Data Accumulation

Data is fed into the machine, which runs it through pre-selected calculations to create a multitude of scenarios based on said data. Through this form of testing, the additive manufacturing machine evaluates the scanning speed, precision, and operational power of the product output for each data point and value.

## Design Calculations

In order to ease the process, the most intuitive notations are applied where possible. First, for the basic equations the pathway in the horizontal pane is denoted as $\chi$ while the time dependent domain, which is occupied by the object that is being produced, is shows as Ω(t) and the build platform itself as Ωpla. As highlighted by Figure 2, the domain is characterized by being a multilayered structure that is built on $\chi$. Meanwhile Γ(t) denotes the external zone which holds the molten metal and eΓ(t) illustrates the part that is heated by the laser. Some further parameters from Figure 2 are the deposition temperature, which can be adjusted based on the laser speed and laser power.

## Basic Equations

As starting point go the calculations the heat conduction equation is used, which is as follows:

$$\forall \boldsymbol{x} \in \Omega(t) \cup \Omega_{\text{pla}},\ \text{div}(\lambda(T)\boldsymbol{\nabla}T(\boldsymbol{x},t)) - \rho c_p(T)\frac{\partial T(\boldsymbol{x},t)}{\partial t} = -\sum_{\phi=1}^{N_\phi} \Delta H_\phi(T)\dot{X}_\phi$$

Within this equation, t is the time; Nφ denotes the amount of phase transitions such as liquid solid or austenite-martensite; cp refers to the respective heat capacity; T equals temperature; ΔHφ refers to the enthalpy change whilst in the φ-th phase

*Figure 2. Notations for calculations.*

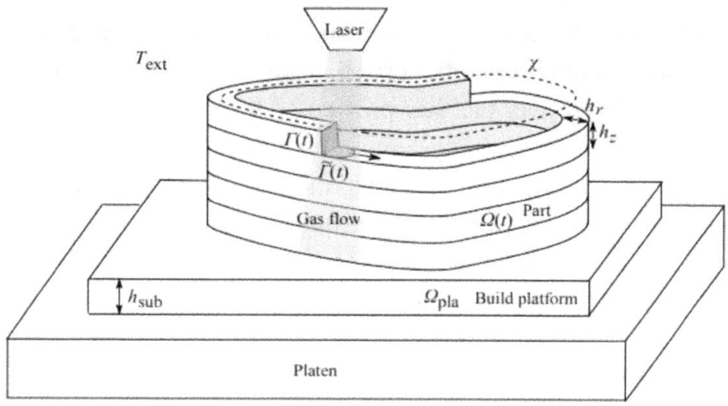


transition; x equals the material point; λ is denotes thermal conductivity; ρ refers to density; and X˙ φ represent the phase proportion rates. For the calculations to make sense, boundary conditions need to be introduced, which include but are not limited to the reduction in heat based on radiation or convection as well as the heat flux, which is applied by the laser and is referred to as qbeam. In summation, these boundary conditions can therefore be summarized as follows:

$$\begin{aligned} &\lambda \nabla T.\boldsymbol{n} = -H(T - T_{ext}) - \sigma\epsilon\left(T^4 - T_{ext}^4\right) && \boldsymbol{x} \in \partial\Omega(t) \cup \partial\Omega_{pla} - \widetilde{\Gamma}(t) - \Gamma(t) \\ &\lambda \nabla T.\boldsymbol{n} = -H(T - T_{ext}) - \sigma\epsilon\left(T^4 - T_{ext}^4\right) + q_{beam} && \boldsymbol{x} \in \widetilde{\Gamma}(t) \\ &T = T_{dep} && \boldsymbol{x} \in \Gamma(t) \end{aligned}$$

With n being the outward normal vector and the initial condition being:

$$q_{beam}(\boldsymbol{x}, t) = \frac{2\eta_{beam} P_{beam}}{\pi R_{beam}^2} \exp\left(-2\frac{\|\boldsymbol{x} - \boldsymbol{x}_{beam}(t)\|^2}{R_{beam}^2}\right)$$

In this equation, Tdep denotes the deposition temperature; σ signifies the Stefan-Boltzmann constant; Text is the room temperature; and H refers to the heat transfer coefficient. For the qbeam variable, a Gaussian model is applied whereby: qbeam(x, t) = 2ηbeamPbeam πR 2 beam exp −2 kx − xbeam(t)k 2 R 2 beam. To this end, ηbeam is the absorption coefficient; Pbeam refers to the laser beam power; xbeam(t) equals the moving heat source location; Rbeam denotes the beam radius; and x refers to the material point of interest. As this shows, for every layer under the flow H is significantly increased by the gas flow.

## Assumptions and Limitations

To start off, the first and third equations introduced are analytically difficult to be solved given the geometrical complexity involved as well as due to the time-dependence of the domain. Furthermore, the heat fluxes aligned with the $\chi$ path are neglected, which is why the successive placements on $\chi$ can be assumed to be independent. Additionally, while pseudo three-dimensional temperature fields can indeed be calculated by applying the previously outlined methodology, these are also restricted because of the geometry involved. On the temperature side there are also some limitations for these calculations as temperatures are large contingencies for phase proportion rates.

## Semi-Analytical Solution

Now that the foundation has been laid, the semi-analytical solution of the heat conduction problem needs to be derived. Given the presence of an alternating algorithm, Q(i), i.e. the volumetric heat, can be assumed to be a known factor for this section before a simple phase transition model is derived to update the Q(i) variable. Subsequently, the mathematical solution is established for the n-th sub-computation ($1 \leq n \leq N$) on the five time interval [tn, tn+1]. Thus, there are n-layers created by the following equation:

$$r, z) \in \left[R_{\text{inf}}, R_{\text{sup}}\right] \times \bigcup_{i=1}^{n} \left[Z^{(i-1)}, Z^{(i)}\right]$$

Whereby Rinf denominates the internal radius; hr = Rsup−Rinf signifies the layer thickness; (i) equals the layer index ($1 \leq i \leq n$); Z (i) are the interfaces positions; Rsup is the external radius; and hz = Z (i)−Z (i−1) indicates the layer height. In order for the equation to work, constant material parameters for the [tn, tn+1] interval are required. For cases in which such material parameters are dependent on temperature, $\lambda$ (i), i.e. the thermal conductivity; and D(i), i.e. the thermal diffusivity, need to be adjusted in the last part of the time interval. This semi-analytical derivation is based on orthogonal projection on a finite dimensional vector space, interpolation of the volumetric heat by exponential fitting, and Fourier-Bessel series expansions.

## Heat Sources, Initial Condition and Boundary Conditions

To overcome this challenge, a negative power per unit volume is added into the equation in order to simulate the heat loss from the gas flow. In summation, the heat conduction equation there reads:

$$\frac{\partial^2 T^{(i)}}{\partial r^2} + \frac{1}{r}\frac{\partial T^{(i)}}{\partial r} + \frac{\partial^2 T^{(i)}}{\partial z^2} - \frac{1}{D^{(i)}}\frac{\partial T^{(i)}}{\partial t} = -\frac{Q^{(i)}(t)}{\lambda^{(i)}}$$

Whereby T(i) denotes the temperature and:

$$Q^{(i)}(t) = \sum_{\phi=1}^{N_\phi} \Delta H_\phi \dot{X}_\phi^{(i)} - \frac{2\sigma\epsilon}{h_r}\left(\left[T^{(i)}\right]^4 - T_{\text{ext}}^4\right) + Q_{\text{beam}}^{(i)}(t) - Q_{\text{gas}}^{(i)}(t)$$

The factor 2 divided by hr (2/hr) is shown as part of the radiactive term in order to change the power at the position of the inner as well as most outward surfaces into a single power-per-unit volume. Moreover, the Q(i) beam denotes the volumetric heat emitted by the laser so that altogether the Gaussian model is derived as follows:

$$Q_{\text{beam}}^{(i)}(t) = \begin{cases} \dfrac{2\eta_{\text{beam}} P_{\text{beam}}}{\pi h_z R_{\text{beam}}^2} \exp\left(-2V_{\text{beam}}^2 \dfrac{(t-t_n)^2}{R_{\text{beam}}^2}\right) \\ 0 \end{cases}$$

In this equation, Vbeam refers to the laser speed while the power-per-unit volume related to the flow of gas, Q(i)gas(t), is related to a convection condition such as, for instance, H(T(i) - Text)). Then, the Gaussian function is used to approximate the heat transfer coefficient related to the gas flow and as a function of the laser speed, which translates to the following equation for Q(i)gas(t):

$$Q_{\text{gas}}^{(i)}(t) = 2\frac{H_{\text{gas}}}{h_r}\left(T^{(i)} - T_{\text{ext}}\right)\exp\left(-2V_{\text{beam}}^2 \frac{(t-t_n)^2}{R_{\text{gas}}^2}\right)$$

Here, Hgas refers to highest level the heat transfer coefficient, which is related to the gas flow, can have. Moreover, Rgas depicts the part that is affected by flow of gas whilst Q(i) gas is dependent on T(i).

## Practical Application: Simulation and Experimentation

The goal of this practical application is to construct a three-dimensional model that can be manufactured as a prototype to an original existing jet engine turbine

*Table 1. Model geometry*

| Bounding Box | |
|---|---|
| Length X | 181.97 mm |
| Length Y | 181.75 mm |
| Length Z | 60. mm |
| **Properties** | |
| Volume | 3.0393e+005 mm³ |
| Mass | 2.3859 kg |
| **Statistics** | |
| Bodies | 31 |
| Active Bodies | 31 |
| Nodes | 118101 |
| Elements | 71535 |
| Mesh Metric | Skewness |
| Min | 5.79276244232485E-04 |
| Max | 0.975240694041518 |
| Average | 0.340631838692623 |
| Standard Deviation | 0.205008198128081 |

and then perform all desired analyses on it. The results will then we used in order to finally create a digital twin to mass-produce the desired component and save the numerous prototyping and analysis costs. To start with, we first need to decide on a component that can be replicated into a downsized ratio and be structured on any CAD software. Here in this book chapter, we have taken the inspiration of our turbine from the GE9X Commercial Aircraft Engine that produces 105,000 pounds of thrust and has 16 turbine blades. This is a 4$^{th}$ generation composite fan blade and the largest engine ever engineered by GE. The GE9X engine claims to have a 10% lower fuel burn compared to the previous GE engine models and has the largest turbine blade diameter in the whole of commercial aviation industry. Due to the claims of efficiency by GE, this model is the perfect base model to establish a digital twin solution that links to the prototypes created for turbines by the means of additive manufacturing. Now that we are familiarized with the concept of our model, we will scale down the model based on assumed sample measurements by GE Aviation with a length unit set to meters, which will give us the X, Y, and Z lengths as 181.97mm, 181.75mm, 60mm respectively (see Table 2).After the model geometry has been set, we will take into consideration the material and its properties used, which for the chosen material for this particular model is structural steel.

Put in practice, the model geometry with its properties and measurements can be visualized in the following figure.

This complete geometrical model can easily be designed using Ansys as the Figures 7-8 above show. To this end, Ansys is used as it is the most suitable comprehensive, scalable software solution that is able to minimize the risk of additive manufacturing processes whilst also ensuring high-quality, certifiable output. By using the Ansys software for additive solutions, it is possible to cover a range from designing for

*Figure 3. Geometrical model of airplane turbine in Ansys.*

ANSYS

## Project

| | |
|---|---|
| First Saved | Monday, June 1, 2020 |
| Last Saved | Monday, June 8, 2020 |
| Product Version | 16.1 Release |
| Save Project Before Solution | No |
| Save Project After Solution | No |

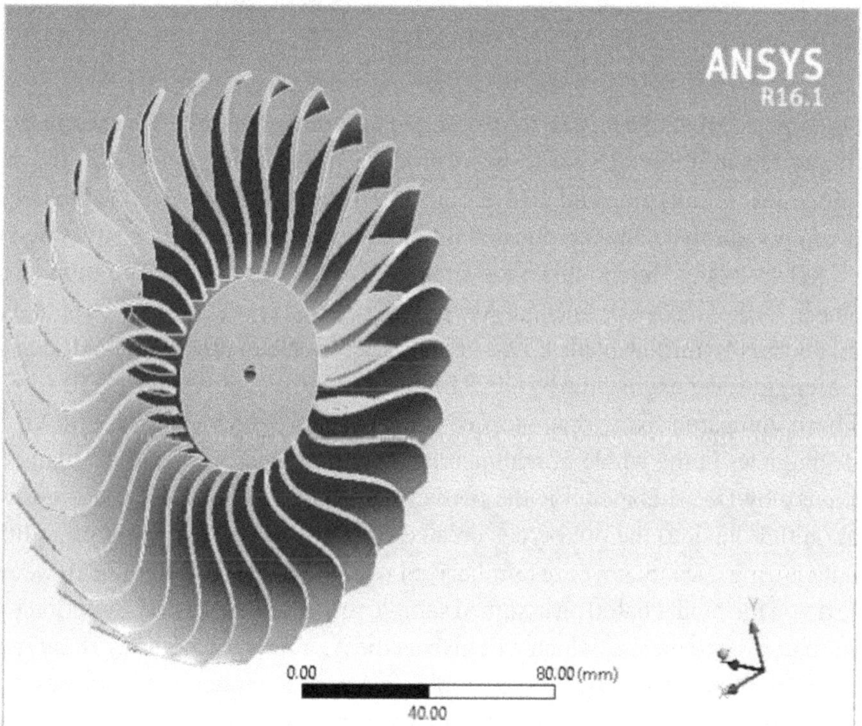


additive manufacturing (DfAM) up to the qualification and certification of parts that have the ability to construct build-file preparations, metal additive manufacturing build process simulation, as well as material analysis at a microstructure level.

*Figure 4. Surface area and parameter of reference plane / turbine head.*

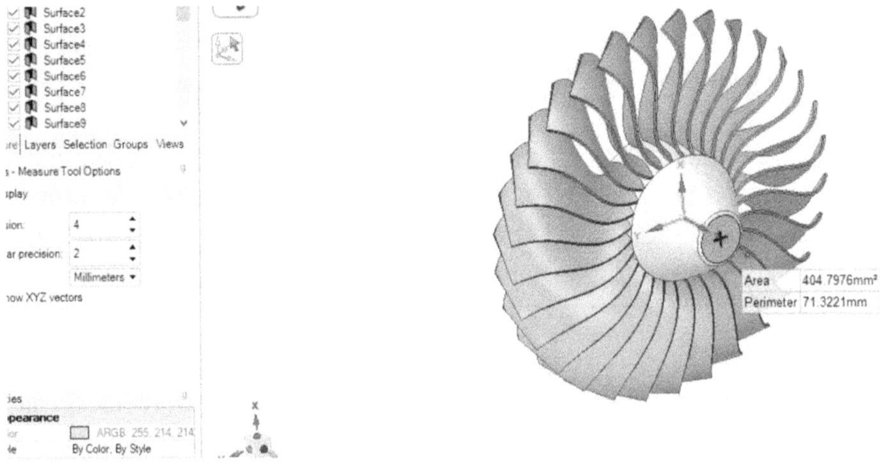


## Structural Analysis Solution

Another advantage of the Ansys software is the fact that it grants access to a complete range of tools for related analyses in order to determine single-load cases together with a vibration or transient analysis. Additionally, it is also possible to analyze (non-) linear behavior of geometry, joints, and materials in general. Ansys also comprises advanced solver technology through the packages of Ansys Autodyn & Ansys LS-DYNA, which allows for the simulation of drops, impact and even explosion. In brief, there are hence 12 possible applications of the Ansys software that make it a compelling case for this book chapter. These applications are: Strength Analysis, Vibration, Thermal Analysis, Durability, Rigid Body Dynamics, Hydrodynamics, Composites, Impact, Optimization, HPC For FEA, Topology Optimization, and, most importantly, Additive Manufacturing.

## Direct Metal Laser Melting Machine (DMLM)

Talking about GE aviation and GE engines as the chosen model, the company GE also manufactures additive manufacturing machines, one of which is the Direct Metal Laser Melting Machine (DMLM) such as the X Line 2000R. The DMLM is a machine that uses the concept of scanning and laser cutting. The scanning is done by the machine to feed the geometric aspect of the work piece. The laser is used to melt the layers of fine metal powder, which when melted are then turned into some of the most complex geometrical designs. Due to the use of laser technology, the final product is incredibly precise to the desired output. GE claims to have the

*Table 2. Input data type for DMLM machine.*

| Scanning speed (mm/s) | Power (W) | Heat input (J/mm) | Cooling rate (K/s) | Growth rate (m/s) | Temp. gradient (K/m) |
|---|---|---|---|---|---|
| 8.3 | 3000 | 361.45 | 167.85 | 1.10E-02 | 1.15E+04 |
| 8.3 | 3375 | 406.63 | 138.30 | 1.29E-02 | 7.97E+03 |
| 8.3 | 3750 | 451.81 | 127.21 | 1.11E-02 | 5.40E+03 |
| 8.3 | 4125 | 496.99 | 117.86 | 1.45E-02 | 4.42E+03 |

largest powder-bed metal additive system globally that is available to provide for the needs of industrial use. The geometry of the work piece can be directly uploaded onto this DMLM machine by the means of nothing but a simple CAD file. The laser power of this machine goes up to 2 x 1000 W (cw) fiber lasers. A sample data set that these machines require is depicted in Table 3 below.

## DISCUSSION OF RESULTS

The curate results clearly show the different desired solutions. The stress was tested in different temperature conditions which were both mapped in order to find out the structural integrity of the product. These conditions were then applied to materials of different properties and structures that were then subjected to all previously mentioned analyses. Due to the fact that, the majority of the results are predominantly mathematical analyses.

As Table 4 above illustrates, the maximum and minimum stress affects the surface of the body and signifies the exact point where the maximum and minimum load occurs. This information allows us then to contemplate the structural properties required to balance out the stress and thermal conditions as indicated by Table 4 below.

The following figures (9,10) all showcase some of the desired analysis by simulation.

*Table 3. Sample table of summary results as retrieved from the Ansys simulation.*

| Results | | | |
|---|---|---|---|
| Minimum | 0. mm | 0. mm/mm | 0. MPa |
| Maximum | 1.976e-002 mm | 4.7912e-006 mm/mm | 0.9447 MPa |
| Minimum Occurs On | Solid | Surface Body | |
| Maximum Occurs On | Surface Body | Solid | |

*Table 4. Summary output of the structural properties required to balance stress and thermal conditions.*

**Structural Steel > Compressive Ultimate Strength**

| Compressive Ultimate Strength MPa |
|---|
| 0 |

**Structural Steel > Compressive Yield Strength**

| Compressive Yield Strength MPa |
|---|
| 250 |

**Structural Steel > Tensile Yield Strength**

| Tensile Yield Strength MPa |
|---|
| 250 |

**Structural Steel > Tensile Ultimate Strength**

| Tensile Ultimate Strength MPa |
|---|
| 460 |

**Structural Steel > Isotropic Secant Coefficient of Thermal Expansion**

| Reference Temperature C |
|---|
| 22 |

The tables above show the different structural properties considered at different temperature gradients that resulted in an overall sufficing result tally. This information is critical as it can vastly impact the wear and tear caused directly onto the airplane turbines. Moreover, it also gives an estimated idea of the shelf life of the turbine since we know that each part that is being assembled within an airplane gets replaced depending on its completed airtime. Knowing the structural integrity and being able to optimize the wear and tear rate, we are now able to bring about a tremendous paradigm shift to the overall performance of the plane regarding its sustainability. Specifically, this process of airplane turbine manufacturing optimization allows us to reduce engine noise as well as the overall environmental footprint of the airplane whilst increasing fuel efficiency.

## CONCLUSION AND FURTHER RESEARCH

Over the last few decades, the demand for air traffic has steadily increased. In addition to safety concerns, increased air travel has a significant negative impact on the environment, both in terms of CO2 emissions and overall anthropogenic

*Figure 5. Total structure deformation.*

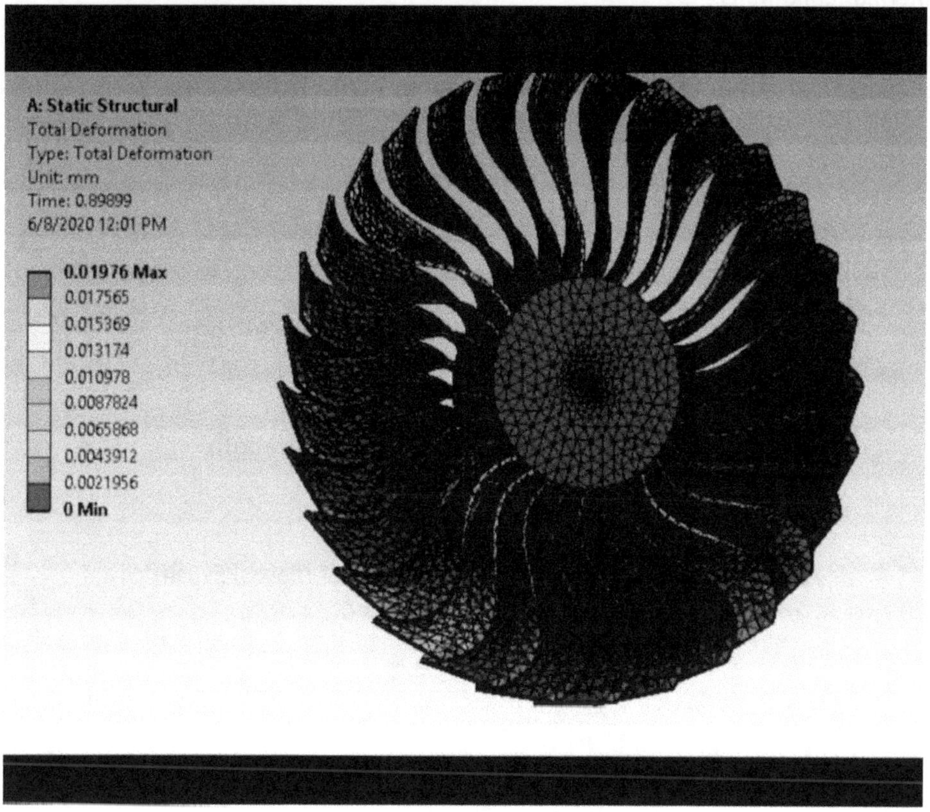


climate change. There has been an increase in interest in research into optimising aeroplane turbines due to their role in these negative developments. To support these efforts, the current book chapter takes a novel and innovative approach by bridging the digital and physical worlds in order to 10x innovation in this area. Because the application of digital twin technology to additive manufacturing for the purpose of optimising product development is still in its infancy, prior research in this field is sparse. Nonetheless, this book chapter performed an end-to-end simulation of a digital twin interface, beginning with data collection and product design and ending with the calculation and testing of semi-analytical solutions before presenting the final digital twin interface. The end result is a significant step toward the sustainability of aeroplane design in particular, but also product development more broadly, because it enables the entire process to be optimised from start to finish in terms of quality, safety, and cost. Further research into the creation of digital twins for additive manufacturing is not only important but highly encouraged. If we are successful in

*Figure 6. Elastic strain on the overall structure.*

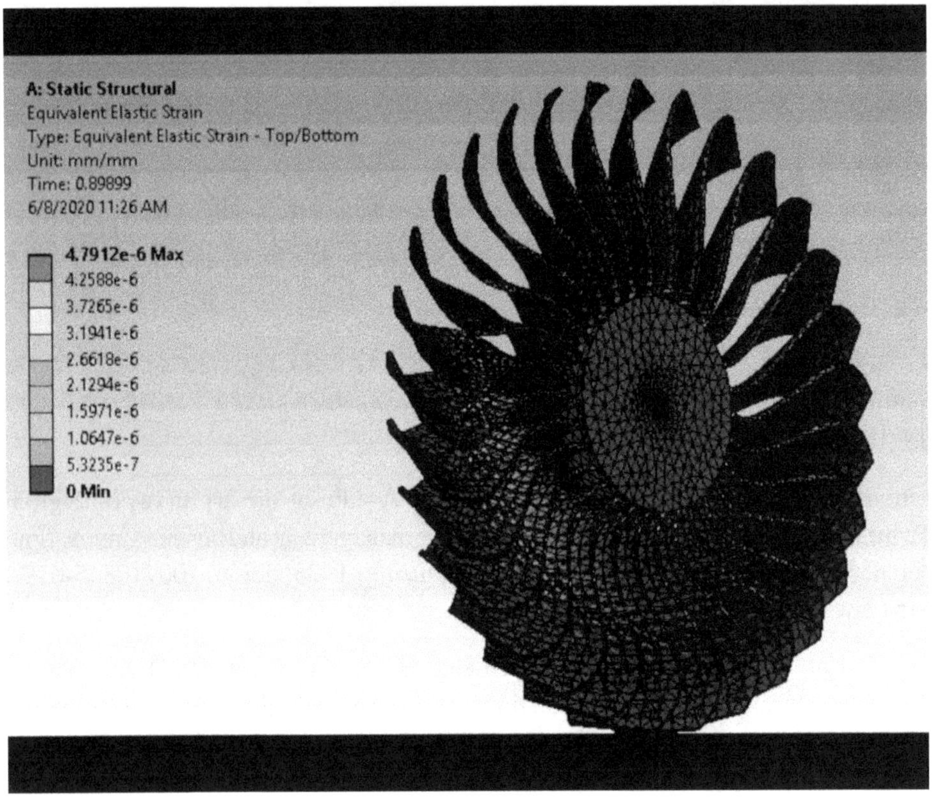


streamlining and improving this process, it can be applied to any industry, resulting in exponential effects in product development on all fronts, regardless of the industry.

## REFERENCES

Aviation Pros. (2012). *Turbine Technology: Innovations in Turbine Engines*. Retrieved July 18, 2020, from https://www.aviationpros.com/aircraft/commercial-airline/article/10617063/turbinetechology-innovations-in-turbine-engines

Aviation Pros. (2013). *Stainless Steel: A metal for all seasons*. Retrieved July 18, 2020, from https://www.aviationpros.com/home/article/10387223/stainless-steel-a-metal-forall-seasons

Aviation Safety Network. (2015). *Grumman G-73T Turbo Mallard*. Retrieved July 18, 2020, from https://aviation-safety.net/database/record.php?id=20051219-0

Boeing. (2019). *Engine Thrust Hazards in the Airport Environment.* Retrieved July 15, 2020, from https://www.boeing.com/commercial/aeromagazine/aero06/textonly/s02txt.html

Debroy, T., Zhang, W., Turner, J., & Babu, S. S. (2017). Building digital twins of 3D printing machines. *Scripta Materialia, 135*, 119–124. doi:10.1016/j.scriptamat.2016.12.005

European Union Aviation Safety Agency. (2019). *Aviation Environmental Impacts.* Retrieved July 6, 2020, from https://www.easa.europa.eu/eaer/climatechange/aviation-environmental-impacts

Körner, C., Attar, E., & Heinl, P. (2011). Mesoscopic simulation of selective beam melting processes. *Journal of Materials Processing Technology, 211*(6), 978–987. doi:10.1016/j.jmatprotec.2010.12.016

Lim, K. Y. H., Zheng, P., & Chen, C. H. (2019). A state-of-the-art survey of Digital Twin: Techniques, engineering product lifecycle management and business innovation perspectives. *Journal of Intelligent Manufacturing*, 1–25. doi:10.100710845-019-01512-w

Lu, Y., Liu, C., Kevin, I., Wang, K., Huang, H., & Xu, X. (2020). Digital Twin-driven smart manufacturing: Connotation, reference model, applications and research issues. *Robotics and Computer-integrated Manufacturing, 61*, 101837. doi:10.1016/j.rcim.2019.101837

Michaleris, P. (2014). Modeling metal deposition in heat transfer analyses of additive manufacturing processes. *Finite Elements in Analysis and Design, 86*, 51–60. doi:10.1016/j.finel.2014.04.003

Ning, J., Sievers, D. E., Garmestani, H., & Liang, S. Y. (2019). Analytical modeling of in-process temperature in powder feed metal additive manufacturing considering heat transfer boundary condition. *International Journal of Precision Engineering and Manufacturing-Green Technology*, 1-9.

Sun, X., Bao, J., Li, J., Zhang, Y., Liu, S., & Zhou, B. (2020). A digital twin-driven approach for the assembly-commissioning of high precision products. *Robotics and Computer-integrated Manufacturing, 61*, 101839. doi:10.1016/j.rcim.2019.101839

Wei, H. L., Mazumder, J., & DebRoy, T. (2015). Evolution of solidification texture during additive manufacturing. *Scientific Reports, 5*(1), 16446. doi:10.1038rep16446 PMID:26553246

Xu, X., Yang, X., Li, J., Pan, S., Bi, Y., & Gao, Y. (2019). Influence of laser energy distribution on laser surface microstructure processing. *Optik (Stuttgart)*, *199*, 163244. doi:10.1016/j.ijleo.2019.163244

# Chapter 9
# Online Detection and Prediction of Fused Deposition Modelled Parts Using Artificial Intelligence

**Sachin Salunkhe**
*Vel Tech Rangarajan Dr. Sagunthala R&D Institute of Science and Technology, India*

**G. Kanagachidambaresan**
*Vel Tech Rangarajan Dr. Sagunthala R&D Institute of Science and Technology, India*

**C. Rajkumar**
*Vel Tech Rangarajan Dr. Sagunthala R&D Institute of Science and Technology, India*

**K. Jayanthi**
*Vel Tech Rangarajan Dr. Sagunthala R&D Institute of Science and Technology, India*

## ABSTRACT

*Fused deposition modelling (FDM) is a technology used for filament deposition of heated plastic filaments by a given pattern by the melted extrusion process. Delamination is a critical issue of FDM's incredibly complex parts. In this chapter, the artificial intelligence (machine learning) model is used for online detections and prediction of FDM parts. The proposed machine learning and convolutional neural network model is capable of online detect delamination of FDM parts. The proposed model can also be applied for different types of additive manufacturing materials with less human interaction.*

DOI: 10.4018/978-1-7998-8516-0.ch009

# INTRODUCTION

## General

Multimaterial, multifunctional designs with complex geometric features with additive manufacturing (AM) have become possible within the last decade. Fused Deposition Modelling (FDM) is one of the most widely used due to its ease of use and low cost. FDM divides CAD models into thin layers of 2D patterns, which are then represented with extruded polymer rasters to create the final product. However, interlayer imperfections still cause delamination and warping, necessitating reprinting and wasting material. A weak adhesion between two layers can lead to delamination, and residual thermal strain from printing can cause warping. Warping can also cause delamination. Imperfections' appearance can be greatly influenced by printing settings such as print parameters, first-layer calibration, and model geometry. Recent advances in the application of artificial intelligence (AI) and machine learning to materials science and engineering problems have enabled researchers to use machine-learning algorithms to classify and predict various printing defections, such as blobs, warps, and delamination. The implementation of a new slicing mechanism to divide prints into spatially locked bricks in order to reduce warping is another intriguing development in the field. Contrary to previous research, which was unable to identify or predict interlayer problems before they spread throughout the entire construction process, this new study does so. This chapter discusses the use of a computer vision and strain gauge model to detect and predict interlayer imperfections in 3D printed parts, such as delamination and warping.

Delamination conditions can be classified and detected using camera-based images and deep learning algorithms. The researchers have also developed a new method of measuring and forecasting future warping based on strain gauge measurements. The first type of interlayer flaw to consider is delamination. Delamination is primarily the result of an incorrect gap between the current nozzle height and the print, which leads to a weak bond between the print's layers. Thus, the literature shows that the first step in resolving this issue is to properly calibrate the nozzle offset value (see Figure 1). Because manual calibration relies so heavily on the user's ability to see the shape of the extruded polymer at the nozzle tip with their naked eye, we've created a system that uses deep learning to mimic manual calibration. An in-house designed and printed camera mount holds a Logitech USB camera to the printing nozzle's left side, allowing the team to monitor the print process in real time. Vibrations caused by printing have been reduced thanks to the mount's cantilevered component being reinforced on the back. As a result of this, the filming angle is nearly horizontal, and the camera's plastic shell has been removed to make room for it in the printing space. Because the offset nozzle height of the first layer remains within the range

of 0.1–0.2 mm, a near horizontal view is preferable to a far horizontal view. Now the distance between the raster's current position is more precisely monitored.

## Defects in 3D Printing

Sensors and feedback controls are only available for the nozzle and build platform temperatures on the vast majority of Fused Filament Fabrication (FFF) 3D printers. Some operational prerequisites, such as filament detection, may require functional check sensors to be present. Commercial (FFF) 3D printers usually don't monitor the printing process or check to see if it's going as smoothly as anticipated. In the event of a defect or even a critical failure, such as the print object losing adhesion to bed, the printing process will continue unless the operator intervenes. This wastes material, power and equipment operation time, and it may also lead to malfunctions in the printing parts that are being used (e.g., nozzle clogging). The result is that print progress must be checked by process operators on a regular basis.

## Delamination in Additive Manufacturing

When a material delaminates, a failure mode occurs in which the layers break down. A variety of materials, such as laminate composites and concrete, can fail due to delamination. Layers in rolled steel, plastics, and 3D-printed metals can separate and fail. It is possible for the coated substrate to separate from any applied surface coatings such as paints or films. The failure of interlayer adhesion is a common occurrence in laminated composites, as the layers separate. There is a weaker polymer matrix connecting the sheets of high strength reinforcement (such as carbon fibre or fibreglass) in fibre reinforced plastics (e.g., epoxy). It is possible for shear loads and loads applied perpendicularly to the high strength layers to lead to polymer matrix fracture or fibre reinforcement debonding. Concrete confines oxidised metal, increasing the volume and increasing the stress on the material. A fracture plane perpendicular to the surface may be created by corroded rebar when stresses exceed the strength of the concrete. Once the fracture plane has formed, the top layer of concrete may come away from the substrate. Figure 1 depicts a typical delamination model in action.

## Wraping in Additive Manufacturing

Warping For deep melt pools, this can occur between two layers or at the boundary between support and part layers (curling). This also occurs when the build is stopped and restarted, and it can occur between two layers or at the boundary between substrate and the first layers defect for deep melt pools.

*Figure 1. Delamination in 3D model (Liu et al., 2019)*

## LITERATURE REVIEW

In Additive Manufacturing, which is also known as 3D printing, powder, wire, or sheets are layered one on top of the other to create parts. When metal components are additively manufactured using laser-powder bed fusion, a Convolutional Neural Network has been trained to detect delamination (Richmond et al., 2017). This helps explain why splatter defects appear. According to the methodology described above, 96.8% of classifications were correct. Cracks, gas porosity, and a lack of fusion in metal additive manufacturing have all been detected with 92.1 percent accuracy using Convolutional Neural Networks. The failure detection rate was 100 percent for the final printed objects and comparisons of differences. (Simonyan & Zisserman, 2014). Clogged nozzles and filament loss were found to be the causes of this issue. To prevent printing from continuing when the reconstructed shape differed by more than 5% from the original STL file, the new method was improved. Using a low-cost 3D printing model and comparing it to a printed shape yielded promising results for detecting effects. FDM is by far the most widely used commercial method. FDM is widely used. However, there are drawbacks to the FDM method, such as missing walls, sagging strends, and misaligned layers (Günaydın & Türkmen, 2018), that must be addressed in addition to the advantages. SLM (Selective Laser Melting)

and selective laser sintering are superior in print quality to FDM (Fused Deposition Modeling) (Ngo et al., 2018).

The goal of computer vision and object detection is to develop techniques and algorithms that allow computers to process, interpret, and comprehend visual data such as video and image (Williams et al., 2019). Developing computer algorithms that could mimic human visual behaviour was a primary goal of early Computer Vision research, which began in the 1960s. Developing computer algorithms that could mimic human visual behaviour was a primary goal of early Computer Vision research, which began in the 1960s. Your efforts should be directed toward optimising delamination by calibrating the nozzle offset height, which is where our observation model has an accuracy of 97.8% when testing and 91.0% when validating. The strain gauge set-up effectively quantifies a print's warping tendency for the purpose of predicting warpage. A warping prediction mechanism can be realised by varying the strain amplitude threshold. Defect detection in laser-powder bed fusion using in-situ thermographic monitoring and a deep learning-based model. (Carbonell et al., 2020) The detection was not based on the printed object's shape, but on the printers' configuration and prior knowledge of their correlations with defects. Consequently, the trained machine detected the printing head's nozzle height from image data and found an association between the detected height and actual defect occurrence. Due to its ease of use and wide range of applications, 3D extrusion printing technology is gaining traction. Reprint 3D printing research is heavily focused on technology, new structural design, and new materials (Jiang et al., 2020). Short carbon thermoplastic material is combined with a continuous carbon filter and used in a variety of engineering applications via 3D printing (Maqsood & Rimašauskas, 2021).

# METHODOLOGY

## Collection and Annotation of Data

The 3D printer has been used to collect images of printed objects that were defective. The images were generated using a straightforward "Stringing test" object that was created. The use of well-known printing parameters, in particular, was found to introduce significant defects. The Single Shot Detector was trained on a final Data set of 500 images showing the stringing phenomenon (Deep Neural Network). Multiple Data Augmentation techniques were used to increase the number of training instances after the training data was gathered for the stringing defect. The term "Data Augmentation" refers to the use of a variety of techniques on existing datasets in order to create new synthetic data that contains all the information necessary in this

case. Each image in the base training set is subjected to half-size scaling (image resize), horizontal flipping, random cropping, 90-degree rotation, and brightness change (randomly).

## Selection and Training of Models

There is a diverse set of existing state-of-the-art algorithms available, each with its own unique set of training, accuracy, and testing speeds on benchmark datasets. The new method must strike a balance between accuracy and detection speed, as the model's purpose is to be deployed in a live environment. The Single Shot Detector was chosen as the model for this chapter. The Single Shot Detector running on a 300 x 300 input (SSD-300) achieved a Mean Average Precision (MAP) of 74.3 percent on the benchmark Dataset VOC-2007 at 59 frames per second (FPS) and a Mean Average Precision (MAP) of 41.2 percent on the benchmark Dataset of Common at an Intersection over Union (IoU) of 0.5. Contextualization of Objects (COCO test-dev2015). While maintaining a high frame rate, the achieved mAP outperformed existing (at the time) state-of-the-art models. Additionally, SSD models offer faster runtime in real-world applications and require less training. Additionally, there is a Single Shot Detector (SSD-512) that operates on 512 x 512 input frames and achieves a higher mAP on VOC-2007 (76.8 percent). SSD-300 was chosen for this new method due to its higher frame rate than SSD-512 (59 vs. 22) and low input resolution requirements, as the model needed to run on low-quality cameras as well.

## Image for Printing

This camera input port on the Raspberry Pi allows users to take high-definition videos and pictures. Python and Raspberry Pi-specific libraries let users build tools that capture images and video in real time and analyse them on the glide, or need them for later processing. This results in a self-contained system that can be used as a tool for item identification, security, or other image processing applications. The objective is to establish the fundamentals of recording video and images onto the Raspberry Pi and analysing them using Python and statistics.

## Convolutional Nueral Network

When it comes to deep learning, convolutional neural networks (also known as ConvNets) are most commonly used to analyse visual imagery. Artificial Neural Networks with Shift Invariance or Space Invariance are other terms for them (SIANN). Input features slide along convolution kernels or filters, which produce translation-equivariant responses known as feature maps based on the shared-

weight architecture. It's a common misconception, but most convolutional neural networks aren't completely invariant in all directions. Computer vision algorithms are used in a wide range of applications including video and image recognition; recommender systems; image classification; and medical image analysis. Multilayer perceptrons are specialised in CNNs. Full-connected networks, in which every neuron in one layer is connected to every neuron in the following layer, are referred to as multilayer perceptrons. Networks that have "complete connectivity" tend to overfit data. Penalizing parameters during training (such as weight decay) or trimming connectivity are common regularisation techniques used to avoid overfitting (skipped connections, dropout, etc.) With CNNs, regularisation is approached in a new way: by embedding smaller and simpler patterns in their filters, they build patterns of increasing complexity.

## Model of a Residential Network

ANNs like residual neural networks (ResNet) are built using known cerebral cortex pyramidal cell structures. Skip connections, or shortcuts, are used by residual neural networks to bypass specific layers. There are two or three layers of skips with nonlinearities between them and batch normalisation in typical ResNet models. An additional weight matrix can be used to learn the skip weights, and these models are known as Highway Nets. Models with multiple parallel skips are known as dense nets. In the context of residual neural networks, a non-residual network is known as a plain network. Reconstruction of a pyramidal cell structure. Axon arbour is labelled blue, while the soma and dendrites are red. The soma, the basal dendrite, the apical dendrite, the axon, and the collateral axon are all components of a neuron's architecture. Skip connections are added for two primary reasons: to prevent gradients from disappearing and to alleviate the Degradation (accuracy saturation) problem, in which adding more layers to a sufficiently deep model increases training error. It is possible that as a result of training, the weights suppress the layer upstream and amplify the layer that was skipped. To keep things simple, only weights associated with the adjacent layer's connection are changed, with no other weights associated with the upstream layer. A single nonlinear layer or all intermediate layers being linear is ideal when using this technique. If this isn't the case, then a weight matrix for the omitted connection must be discovered (a Highway Net should be used).

## Function of Softmax

Generalization of the logistic function on multiple dimensions is called the softmax function. It is also known as the softargmax 184 or normalised exponential 198. To normalise the network's output to a probability distribution over predicted output

classes, it is frequently used as the neural network's final activation function. To normalise a vector of K real numbers, the softmax function takes an input vector z and normalises it to a probability distribution with K probabilities proportional to exponentials of the input numbers. To put it another way, some vector components may be negative or greater than one before applying softmax and may not add to one; however, after applying softmax, each component will be in the thin interval and the components will add to one, allowing them to be interpreted as probabilities. Greater probabilities are associated with larger input components.

### Super Resolution Image

An imaging system's resolution can be improved using a group of techniques known as super-resolution imaging (SR). In optical SR, the diffraction limit of systems is overcome, while in geometrical SR, the resolution of digital imaging sensors is increased. There are methods based on subspace decomposition (e.g. MUSIC(Liu et al., 2019)) and compressed sensing (e.g. SAMV(Sames et al., 2016)) and algorithms based on compressed sensing (e.g. MUSIC) that can achieve SR over the standard period gramme algorithm in certain radar and sonar imaging applications. Both general image processing and super-resolution microscopy make use of techniques for high-resolution imaging.

## MATERIALS

### The Materials Utilized in 3D Printing

It's as diverse as the finished products when it comes to 3D printing materials. As a result, 3D printing is flexible enough to let manufacturers customise the look, feel, and strength of their products. Best of all, compared to traditional manufacturing processes, these features can be achieved in far fewer steps. A wide range of 3D printing materials can be used to create these products. To turn a 3D print into a finished product, you must first send the printer a high-resolution image of your design. It's all done in Standard Triangle Language (STL), which allows a computerised 3D printer to see the design from all sides and angles because of its complexities and dimensions. The following materials are frequently used: i. Powder ii. Resins iii. Metal iv. Carbon Fiber v. Graphite and Graphene

*Figure 2. Methodology*

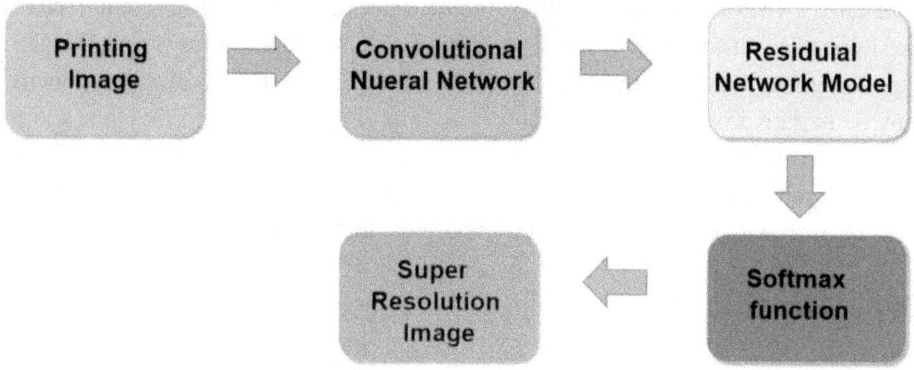


## Polyastic Acid (PLA)

Polyastic acid is a biodegradable 3D printing material made from natural sources like sugar cane and corn starch. 3D printing will be dominated by plastics made from polyastic acid in the near future. It is possible to get them in soft or hard versions. As a result, hard PLA is better suited for use in a wider range of products. PLA's properties are listed in Table 1.

## Components of Printing Image

### Raspberry Pi Zero W Development Board with Case of 4GB

Raspberry Pi is a microcomputer with a single integrated circuit (Figure 3). It is being developed in the United Kingdom by the Raspberry Pi Foundation in collaboration with Broadcom. Raspberry Pi Zero is the smallest chipset in the Raspberry Pi series, at 40% faster but nearly half the size of the original Raspberry Pi.

Prior to powering on the Pi Zero, the SD card must be programmed with an operating system. Similarly, to how a computer requires Windows, Mac OS X, or Linux to function, the Raspberry Pi requires something to help it boot and run software. In addition to the port shown in Figure 4 and 5, all current models of Raspberry Pi include a port for connecting the Camera Module.

*Table 1. Properties of PLA*

| PROPERTIES | | | ANNEALED | |
|---|---|---|---|---|
| | | PLLA | PLLA | PDLLA |
| Tensile strength | Mpa | 59 | 66 | 44 |
| Elongation at break | % | 7 | 4 | 5.4 |
| Modulus of elasticity | MPa | 3750 | 4150 | 3900 |
| Yield strength | MPa | 70 | 70 | 53 |
| Flexural strength | MPa | 106 | 119 | 88 |
| Unnotched izod impact | J/m | 195 | 350 | 150 |
| Notched izod impact | J/m | 26 | 66 | 18 |
| Rockwell hardness | °C | 88 | 61 | 76 |
| Heat Deflection Temperature | °C | 59 | 165 | 52 |
| Vicat Penetration | °C | 88 | 88 | 76 |

*Figure 3. Raspberry pi zero*

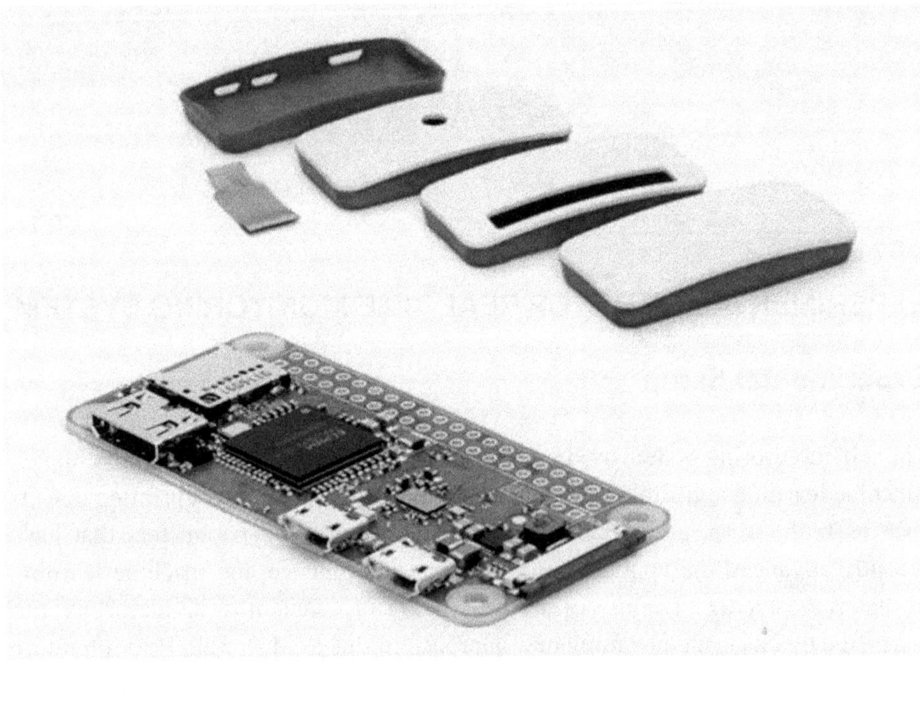

*Figure 4. Raspberry pi camera port*

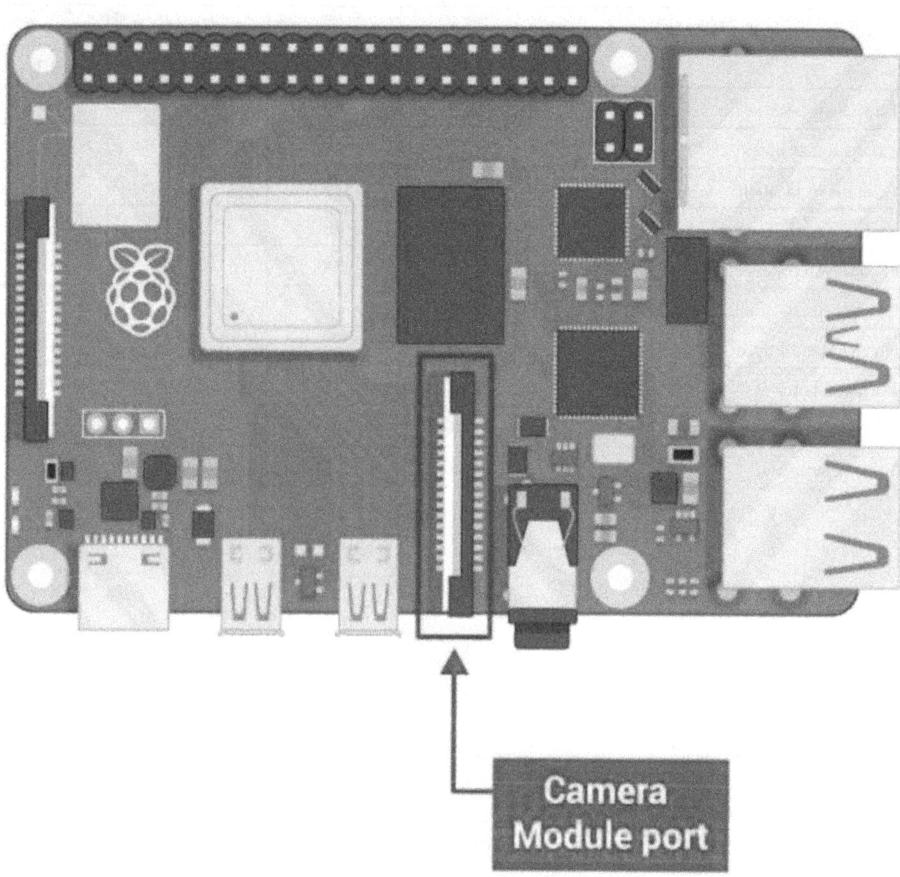


# EXPERIMENTAL SETUP OF REAL TIME MONITORING SYSTEM

## Experimental Setup

The self-monitoring system used in this process is discussed in detail in this chapter. Machine learning algorithms are being used to classify and predict printing defects such as blob, warp, and delamination in light of printing parameters that have recently advanced the application of artificial intelligence and machine learning to materials science. The use of a novel slicing mechanism to divide prints into warp-free blocks is an intriguing new approach in this field as well. Prior literature research, on the other hand, is unable to analyse or forecast interlayer difficulties in real time before they spread to the rest of the construction section. One new way to detect and predict interlayer flaws like delamination and warping in 3D-printed

*Figure 5. Raspberry pi kit together*

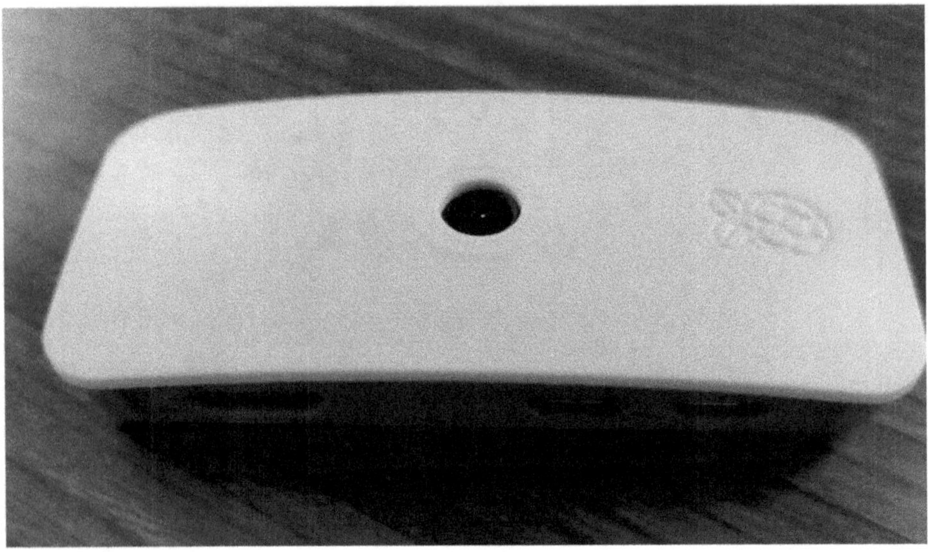

parts is described in this article. The new approach uses camera-based images and deep learning algorithms to classify and detect delamination conditions, and a novel strain gauge measurement setup is used to evaluate and anticipate warping.

## Positioning of the Camera

A custom-made camera mount holds a USB RASPBERRY NOIR camera on the left side of the printing nozzle. The mount's cantilevered component is reinforced to reduce vibration during printing. The camera's plastic shell is also removed to fit within the camera's limited printing space, as shown in Figure 6.

- 8 Megapixel camera capable of 32802464 pixel photographs.
- Capture video in the following resolutions: 1080p30, 720p60, and 640x480p90.
- All software is supported by the most recent version of the Raspbian operating system.

- 220 degrees Fahrenheit at ambient temperature. No heating plate is recommended for the platform. The Flash forge finder 3D printer is completely safe for children, adults, and educators. It is designed to print exclusively with PLA and PLA-added filament.

*Figure 6. Positioning of camera module*

- There is no heat plate. There are no visible wires. Certifications CE, fcc, and ISO 9001 make operation simple for anyone. Simple built-in filament design, with an alarm that sounds when the filament is depleted.
- Professional single extruder structure, virtually eliminates clogs. Leveling of assistance. Assure flawless file transfers and avoid interrupt risk specifications with 4g storage.
- Dimensionsofthepackage:505x500x545mm.Sizeoffinder:420x420x420mm. 0.1-0.2mm resolution. Volume of construction: 140x140x140mm. Pla. Filament. Stl, obj, g file formats.
- Windows xp, windows vista, windows 7/8, and mac os x 3.5-inch touch screen with 16 languages. Flashprint's fundamental parameter. One technology is used in the extruder: fused deposition modelling.
- Positioning accuracy: 0. 0025mm on the Z axis, 0.011mm on the xy axis. Accuracy of printing: 0.1mm. 0.1-0.5mm layer thickness 0.4mm diameter of the nozzle. Shaft speed ranges between 40 and 200mm/s.
- Flow rate of the nozzle: 24cc/h. Suggestions for extr

## Delamination Occuring Images

Aspects such as cost, shape of the object, texture of the object, and utility of the product will be taken into consideration when determining the standards for the material. The delamination of FDM parts is depicted in Figure 7.

*Figure 7. Delaminated product while printing from side view*

When a material fails, it delaminates, which means it fractures into multiple layers. Delamination can occur in a wide range of materials, including laminate composites and concrete, among others. Processes such as rolling and 3D printing can create layers in materials such as steel, which can fail due to layer separation. Other materials such as plastics and metals can also fail due to layer separation. The delamination of printing parts from top view is shown in Figure 8.

## TRAINING RESULTS OF CNN ALGORITHM

Precision was 0.44 and recall was 0.69 with an IoU of 0.4; Precision was 0.41 and recall was 0.63 with an IoU of 0.5; and Recall was 0.62 with an IoU of 0.6, according to the trained model. Predictions must become more precise as the IoU threshold is raised, resulting in metrics that are less accurate. Increase the threshold for approving the classifier's guess may lead to an increase in False Positives, however (lower precision). As long as the predicted defect overlaps at least 40% with any given true defect, it should be treated as a true positive and handled accordingly. A threshold

*Figure 8. Delaminated product while printing from top view*

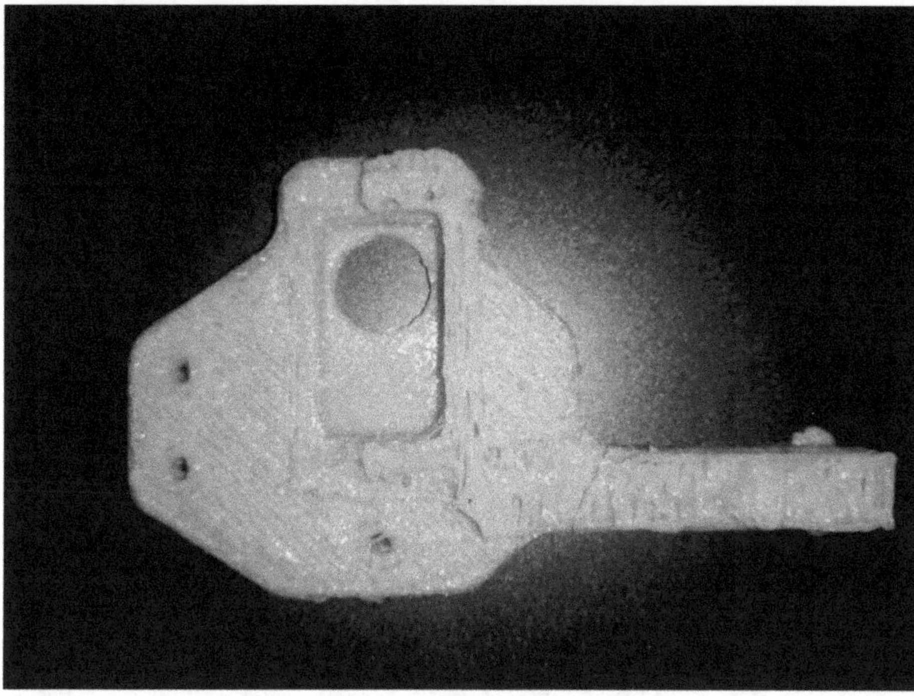

of 0.4 is acceptable in our case. We must keep the IloU threshold as low as possible to increase the number of detected true defects (True Positives) using the current training data. When the IoU threshold is less than 0.4, significant false detections occur (False Positives). As a result, raising the IoU cutoff will result in fewer True Positives while also reducing the number of False Positives that are detected. It's a matter of weighing these two factors when determining the IoU threshold value. If you use this new method (defect detection), you may be alarmed erroneously. False Positives (normalities mistaken for defects) are expensive, so Precision must be high.

## CONCLUSION

The proposed model, which is based on a machine learning model, detects defects such as delamination online monitoring system. The efforts put forth in the proposed model are used to calibrate the nozzle height. It was found that the developed model is capable of predicting accuracy levels of approximately 98%. Additionally, the online detection model that has been proposed can be applied to other 3D printing systems, resulting in improved final product quality with no need for human intervention.

## REFERENCES

Carbonell, M., Fornés, A., Villegas, M., & Lladós, J. (2020). A neural model for text localization, transcription and named entity recognition in full pages. *Pattern Recognition Letters*, *136*, 219–227. doi:10.1016/j.patrec.2020.05.001

Günaydın, K., & Türkmen, H. S. (2018). Common FDM 3D printing defects. *International Congress on 3D Printing (Additive Manufacturing) Technologies and Digital Industry.*

Jiang, Z., Diggle, B., Tan, M. L., Viktorova, J., Bennett, C. W., & Connal, L. A. (2020). Extrusion 3D printing of polymeric materials with advanced properties. *Advancement of Science*, *7*(17), 2001379. doi:10.1002/advs.202001379 PMID:32999820

Liu, Z., Lei, Q., & Xing, S. (2019). Mechanical characteristics of wood, ceramic, metal and carbon fiber-based PLA composites fabricated by FDM. *Journal of Materials Research and Technology*, *8*(5), 3741–3751. doi:10.1016/j.jmrt.2019.06.034

Maqsood, N., & Rimašauskas, M. (2021). Delamination observation occurred during the flexural bending in additively manufactured PLA-short carbon fiber filament reinforced with continuous carbon fiber composite. *Results in Engineering*, *11*, 100246. doi:10.1016/j.rineng.2021.100246

Ngo, T. D., Kashani, A., Imbalzano, G., Nguyen, K. T., & Hui, D. (2018). Additive manufacturing (3D printing): A review of materials, methods, applications and challenges. *Composites. Part B, Engineering*, *143*, 172–196. doi:10.1016/j.compositesb.2018.02.012

Richmond, Libonati, & Wettermark. (2017). Detector. *Proceedings of the European Conference on Computer Vision.* . doi:10.6084/m9.figshare.12221543

Sames, W. J., List, F. A., Pannala, S., Dehoff, R. R., & Babu, S. S. (2016). The metallurgy and processing science of metal additive manufacturing. *International Materials Reviews*, *61*(5), 315–360. doi:10.1080/09506608.2015.1116649

Simonyan, K., & Zisserman, A. (2014). *Very deep convolutional networks for large-scale image recognition.* http://purl.fcla.edu/fcla/etd/CFE0006397

Williams, C.K.L, Winn, J., & Zisserman, F. (2019). The Pascal Visual Object Classes (VOC) Challenge. International Journal of Computer. *Vision (Basel)*, *88*, 303–338.

# Conclusion

At the moment, additive manufacturing processes are constrained by geometrical characteristics, material type, and time constraints. Numerous researchers have been examining the suitability of a product for 3D printing. The researchers developed a printability algorithm that determines whether a component should be manufactured using 3D printing or traditional methods based on a variety of factors such as feature, material, and time. There is potential to expand 3D printing's applicability to other materials as well. The slicer algorithm generates tool paths that are very similar to those generated by a CNC machine programme. The slicing algorithm specifies the path that a 3D printer's head should follow while manufacturing a product. The efficiency of 3D printing is highly dependent on the slicing algorithm. Numerous researchers have been working to optimise the slicing algorithm's efficiency. Further research on this subject is extremely worthwhile. Components with overhang require support, resulting in wastes and increased manufacturing costs in the additive manufacturing. Many researchers have investigated the relationship between component orientation and support volume requirements. When it came to minimising material waste, they had used artificial intelligence-based techniques. This area of AM is still in its infancy, and additional research will be required in the future. The process of additive manufacturing consumes more energy than conventional machining. As a result, additive manufacturing processes are inefficient in terms of energy. As a result, additional research is required to improve the energy efficiency of additive manufacturing. Additionally, this would reduce manufacturing costs. Additive manufacturing would be expanded. During the manufacturing process of a product, the additive manufacturing process generates billions of small particles, which may be harmful if inhaled by the consumer. Numerous researchers have proposed methods for predictive maintenance of machine tools that are based on machine learning. These methods would aid in accurately predicting a machine tool's remaining useful life. This would assist in decreasing the amount of time the machine is down. This would also aid in estimating the quantity of spare parts required. Thus, spare parts would be prepared prior to the machine failing, reducing the amount of time the machine would be offline. The application of artificial intelligence in additive

manufacturing is still in its infancy, and more research is needed in the areas of (i) reducing energy consumption by 3d printers, (ii) expanding productivity during additive manufacturing, (iii) increasing the use of AI in additive manufacturing for cost reduction, (iv) increasing accuracy in predicting the remaining life of machine tools, and (v) minimising defects during additive manufactured products.

# Compilation of References

Boschetto, Giordano, & Veniali. (2013). Surface roughness prediction in fused deposition modelling by neural networks. *The International Journal of Advanced Manufacturing Technology.*

Calignano, F., Manfredi, D., Ambrosio, E. P., Iuliano, L., & Fino, P. (2012). Influence of process parameters on surface roughness of aluminum parts produced by DMLS. *International Journal of Advanced Manufacturing Technology*, *67*(9-12), 2743–2751. doi:10.100700170-012-4688-9

Maqsood, N., & Rimašauskas, M. (2021). Delamination observation occurred during the flexural bending in additively manufactured PLA-short carbon fiber filament reinforced with continuous carbon fiber composite. *Results in Engineering*, *11*, 100246. doi:10.1016/j.rineng.2021.100246

Ning, Cong, Qiu, Wei, & Wang. (2015). Additive Manufacturing of carbon fiber reinforced thermoplastic composites using fused deposition modeling. *Composites Part B, 80*, 369-378.

Raghavendra, N., Koranne, R., Pal, S., Pal, S. K., & Samantaray, A. K. (2009). Joint Strength Prediction in a Pulsed MIG Welding Process Using Hybrid Neuro Ant Colony-Optimized Model. *International Journal of Advanced Manufacturing Technology*, *41*(7–8), 694–705. doi:10.100700170-008-1517-2

Sinha, R. P., & Häder, D.-P. (2002). UV-induced DNA damage and repair: A review. *Photochemical & Photobiological Sciences*, *1*(4), 225–236. doi:10.1039/b201230h PMID:12661961

Ahn, D., Kweon, J.-H., Choi, J.-H., & Lee, S.-H. (2011). Relation between surface roughness and overlap interval in Fused Deposition Modeling. *Advanced Materials Research*, *264-265*, 1625–1630. doi:10.4028/www.scientific.net/AMR.264-265.1625

Barakat, Abdel Ghany, Abd Elsalam, & El Atriby. (2018). Modeling of the Weld Bead Shape at Different Welding Parameters. *AMME-18 The 9th International Scientific Conference of The Military Technical College*, 3–5.

Gholaminezhad, I., Assimi, H., Jamali, A., & Vajari, D. A. (2016). Uncertainty quantification and robust modeling of selective laser melting process using stochastic multi-objective approach. *International Journal of Advanced Manufacturing Technology*, *86*(5-8), 1425–1441. doi:10.100700170-015-8238-0

Hull, C. W. (1984). *Apparatus for production of three-dimensional objects by stereolithography.* United States Patent, Appl., No. 638905.

Jaya Christiyan, K. G., Chandrasekhar, U., & Venkateswarlu, K. (2016). A study on the influence of process parameters on the Mechanical Properties of 3D printed ABS composite. *IOP Conf. Series: Materials Science and Engineering, 114.* 10.1088/1757-899X/114/1/012109

Mahapatra, S. S., & Sood, A. K. (2011). Bayesian regularization-based Levenberg–Marquardt neural model combined with BFOA for improving surface finish of FDM processed part. *International Journal of Advanced Manufacturing Technology, 60*(9-12), 1223–1235. doi:10.100700170-011-3675-x

Melchels, F. P., Feijen, J., & Grijpma, D. W. (2010). A review on stereolithography and its applications in biomedical engineering. *Biomaterials, 31*(24), 6121–6130. doi:10.1016/j.biomaterials.2010.04.050 PMID:20478613

Phatak, A. M., & Pande, S. S. (2012). Optimum part orientation in rapid prototyping using genetic algorithm. *Journal of Manufacturing Systems, 31*(4), 395–402. doi:10.1016/j.jmsy.2012.07.001

Ramadan, A., Barakat, A., & Elatriby, S. (2018). Optimizing Ant Colony Using Harmony Search Optimization. *4th International Conference Welding and Failure Analysis of Engineering, Materials*, 3–9. 10.13140/RG.2.2.10327.32161

Boschetto, V., Giordano, V., & Veniali, F. (2013). Surface roughness prediction in fused deposition modelling by neural networks. *International Journal of Advanced Manufacturing Technology, 67*(9-12), 2727–2742. doi:10.100700170-012-4687-x

Pham, D. T., & Gault, R. S. (1998). A comparison of rapid prototyping technologies. *International Journal of Machine Tools & Manufacture, 38*(10-11), 1257–1287. doi:10.1016/S0890-6955(97)00137-5

Strano, G., Hao, L., Everson, R. M., & Evans, K. E. (2011). Multi-objective optimization of selective laser sintering processes for surface quality and energy saving. *Proceedings of the Institution of Mechanical Engineers. Part B, Journal of Engineering Manufacture, 225*(9), 1673–1682. doi:10.1177/0954405411402925

Bártolo, P. J. (2011). *Stereolithography: materials, processes and applications.* Springer Science & Business Media. doi:10.1007/978-0-387-92904-0

Javed, S., Mahmoudi, A., & Khan, A. M. (2018). Investigation of drilling parameters on hybrid polymer composites using grey relational analysis, regression, fuzzy logic, and ANN models: A critical note. *Journal of the Brazilian Society of Mechanical Sciences and Engineering, 40*(12), 560. doi:10.100740430-018-1470-4

Wu, H., Wang, Y., & Yu, Z. (2015). In Situ Monitoring of FDM Machine Condition via Acoustic Emission. *International Journal of Advanced Manufacturing Technology.* Advance online publication. doi:10.100700170-015-7809-4

Peng, A., Xiao, X., & Yue, R. (2014). Process parameter optimization for fused deposition modeling using response surface methodology combined with fuzzy inference system. *International Journal of Advanced Manufacturing Technology*, *73*(1-4), 87–100. doi:10.100700170-014-5796-5

Raman, R., & Bashir, R. (2015). Stereolithographic 3D bioprinting for biomedical applications. In Essentials of 3D Biofabrication and Translation. Elsevier. doi:10.1016/B978-0-12-800972-7.00006-2

Wu, H., Yu, Z., & Wang, Y. (2017). Real-Time FDM Machine Condition Monitoring and Diagnosis Based on Acoustic Emission and Hidden Semi-Markov Model. *International Journal of Advanced Manufacturing Technology*, *90*(5–8), 2027–2036. doi:10.100700170-016-9548-6

Johnson, D. W., Sherborne, C., Didsbury, M. P., Pateman, C., Cameron, N. R., & Claeyssens, F. (2013). Macrostructuring of emulsion-templated porous polymers by 3D laser patterning. *Advanced Materials*, *25*(23), 3178–3181. doi:10.1002/adma.201300552 PMID:23606519

Liu, C., Roberson, D., & Kong, Z. (2017). Textural Analysis-Based Online Closed-Loop Quality Control for Additive Manufacturing Processes. *2017 Industrial and Systems Engineering Conference*, Blacksburg, VA.

Panda, B. N., Bahubalendruni, M. V. A. R., & Biswal, B. B. (2014). A general regression neural network approach for the evaluation of compressive strength of FDM prototypes. *Neural Computing & Applications*, *26*(5), 1129–1136. doi:10.100700521-014-1788-5

Garg, A., & Tai, K. (2014). An Ensemble Approach of Machine Learning in Evaluation of Mechanical Property of the Rapid Prototyping Fabricated Prototype. *Applied Mechanics and Materials*, *575*, 493–496. doi:10.4028/www.scientific.net/AMM.575.493

Vijayaraghavan, V., Garg, A., Lam, J. S. L., Panda, B., & Mahapatra, S. S. (2014). Process characterisation of 3D-printed FDM components using improved evolutionary computational approach. *International Journal of Advanced Manufacturing Technology*, *78*(5-8), 781–793. doi:10.100700170-014-6679-5

Weiß, T. (2009). Two-photon polymerization for microfabrication of three-dimensional scaffolds for tissue engineering application. *Engineering in Life Sciences*, *9*(5), 384–390. doi:10.1002/elsc.200900002

Spangenberg, A. (2013). Recent advances in two-photon stereolithography. *Updates in Advanced Lithography*, 35-63.

Zhu, Z., Anwer, N., Huang, Q., & Mathieu, L. (2018). Machine learning in tolerancing for additive manufacturing. *CIRP Ann. Manuf. Technol.*, *67*(1), 157–160. doi:10.1016/j.cirp.2018.04.119

Sood, A. K., Ohdar, R. K., & Mahapatra, S. S. (2009). Parametric appraisal of fused deposition modelling process using the grey Taguchi method. *Proceedings of the Institution of Mechanical Engineers. Part B, Journal of Engineering Manufacture*, *224*(1), 135–145. doi:10.1243/09544054JEM1565

Yia, Gläßnera, & Krenkela. (2019). *Energy simulation of the fused deposition modeling process using machine learning approach*. Academic Press.

Ziemian & Crawn. (2001). *Computer aided decision support for fused deposition modeling*. Academic Press.

Kandananond, K. (2020). Optimization of fused filament fabrication system by response surface method. *International Journal of Metrology and Quality Engineering*, *11*, 4. doi:10.1051/ijmqe/2020002

Liu, Z., Lei, Q., & Xing, S. (2019). Mechanical characteristics of wood, ceramic, metal and carbon fiber-based PLA composites fabricated by FDM. *Journal of Materials Research and Technology*, *8*(5), 3741–3751. doi:10.1016/j.jmrt.2019.06.034

Rajamani, D., & Balasubramanian, E. (2019a). Investigations of sintering parameters on viscoelastic behaviour of selective heat sintered HDPE parts. *Journal of Applied Science and Engineering*, *22*(3), 391–402.

Razvi, S. S., Feng, S., Narayanan, A., Lee, Y.-T. T., & Witherell, P. (2019). A Review of Machine Learning Applications in Additive Manufacturing. *Proceedings of the ASME 2019 International Design Engineering Technical Conferences and Computers and Information in Engineering Conference IDETC/CIE2019*. 10.1115/DETC2019-98415

Salonitis, K. (2016). Design for Additive Manufacturing Based on the Axiomatic Design Method. *International Journal of Advanced Manufacturing Technology*, *87*(1–4), 989–996. doi:10.100700170-016-8540-5

Schubert, Van Langeveld, & Donoso. (2014). *Innovations in 3D printing: a 3D overview from optics to organs*. Academic Press.

Chua, C. K., Leong, K. F., & Lim, C. S. (2010). *Rapid prototyping: principles and applications (with companion CD-ROM)*. World Scientific Publishing Company. doi:10.1142/6665

Tamilarasan, A., & Rajamani, D. (2017). Multi-response optimization of Nd: YAG laser cutting parameters of Ti-6Al-4V superalloy sheet. *Journal of Mechanical Science and Technology*, *31*(2), 813–821. doi:10.100712206-017-0133-1

Thrimurthulu, K., Pandey, P. M., & Reddy, N. V. (2004). Optimum part deposition orientation in fused deposition modeling. *International Journal of Machine Tools & Manufacture*, *44*(6), 585–594. doi:10.1016/j.ijmachtools.2003.12.004

Ananthakumar, K., Rajamani, D., Balasubramanian, E., & Paulo Davim, J. (2019). Measurement and optimization of multi-response characteristics in plasma arc cutting of Monel 400™ using RSM and TOPSIS. *Measurement*, *135*, 725–737. doi:10.1016/j.measurement.2018.12.010

Kruth, J.-P. (2003). *Lasers and materials in selective laser sintering*. Academic Press.

Smith, J., Xiong, W., Yan, W., Lin, S., Cheng, P., Kafka, O. L., Wagner, G. J., Cao, J., & Liu, W. K. (2016). Linking Process, Structure, Property, and Performance for Metal-Based Additive Manufacturing: Computational Approaches with Experimental Support. *Computational Mechanics*, *57*(4), 583–610. doi:10.100700466-015-1240-4

Kruth, J.-P. (2007). *Consolidation phenomena in laser and powder-bed based layered manufacturing*. Academic Press.

Rajamani, D., Ananthakumar, K., Balasubramanian, E., & Paulo Davim, J. (2018). Experimental investigation and optimization of PAC parameters on Monel 400™ superalloy. *Materials and Manufacturing Processes*, *33*(16), 1864–1873. doi:10.1080/10426914.2018.1532085

Mirjalili, S. (2015). Dragonfly algorithm: A new meta-heuristic optimization technique for solving single-objective, discrete, and multi-objective problems. *Neural Computing & Applications*, *27*(4), 1053–1073. doi:10.100700521-015-1920-1

Wohlers, T. T. (2009). *Wohlers report 2009: state of the industry: annual worldwide progress report*. Wohlers Associates.

Esakki, B., Ali, T. K., Rajamani, D., & Sachin, S. (2021). Parametric Optimization on Impact Strength of Selective Inhibition Sintering Fabricated PA-12 Parts Based on Evolutionary Optimization Algorithms. *Journal of Materials Engineering and Performance*, *30*(7), 5356–5367. Advance online publication. doi:10.100711665-021-05794-1

Yadroitsev, I. (2007). *Strategy of manufacturing components with designed internal structure by selective laser melting of metallic powder*. Academic Press.

Ciurana, J., Arias, G., & Ozel, T. (2009). Neural Network Modeling and Particle Swarm Optimization (PSO) of Process Parameters in Pulsed Laser Micromachining of Hardened AISI H13 Steel. *Materials and Manufacturing Processes*, *24*(3), 358–368. doi:10.1080/10426910802679568

Rochus, P. (2007). *New applications of rapid prototyping and rapid manufacturing (RP/RM) technologies for space instrumentation*. Academic Press.

Levy, Schindel, & Kruth. (2003). *Rapid manufacturing and rapid tooling with layer manufacturing (LM) technologies, state of the art and future perspectives*. Academic Press.

Cheng & Huang. (2020). *Preparation and characterization of color photocurable resins for full-color material jetting additive manufacturing*. Academic Press.

Gupta & Gurumoorthy. (2021). *Feature-based ontological framework for semantic interoperability in product development*. Academic Press.

Gupta & Gurumoorthy. (2008). *Extending feature based semantic interoperability for sheet-metal product development*. Academic Press.

Bandyopadhyay, A., Bose, S., & Das, S. (2015). 3D printing of biomaterials. *MRS Bulletin*, *40*(2), 108–115. doi:10.1557/mrs.2015.3

Hamoud, M. (2018a). Enhancement of Sliced Layer Contour for Reconstructing A 3D Model Using Solid Freeform Fabrication. *18th International Conference on Applied Mechanics and Mechanical Engineering-AMME-18*, 196-208. 10.21608/amme.2018.35012

Rajamani, D., & Balasubramanian, E. (2019b). Effects of heat energy on morphology and properties of selective inhibition sintered high density polyethylene. *Journal of Mechanical Engineering Science, 13*(1), 4403–4414. doi:10.15282/jmes.13.1.2019.05.0375

Sames, W. J., List, F. A., Pannala, S., Dehoff, R. R., & Babu, S. S. (2016). The metallurgy and processing science of metal additive manufacturing. *International Materials Reviews, 61*(5), 315–360. doi:10.1080/09506608.2015.1116649

Sumesh, A., Rameshkumar, K., Raja, A., Mohandas, K., Santhakumari, A., & Shyambabu, R. (2017). Establishing Correlation Between Current and Voltage Signatures of the Arc and Weld Defects in GMAW Process. *Arabian Journal for Science and Engineering, 42*(11), 4649–4665. doi:10.100713369-017-2609-9

Witherell, P. (2018). Emerging Datasets and Analytics Opportunities in Metals Additive Manufacturing. *Direct Digital Manufacturing Conference*, Berlin, Germany.

Nalluri, S. (1994). *Form feature generating model for feature technology*. Indian Institute of Science, Department of Mechanical Engineering Bangalore.

Nyirenda & Bronsvoort. (2009). *A framework for extendable freeform surface feature modelling*. Academic Press.

Nyirenda, P.J. (2005). *A generic taxonomy for defining freeform feature classes*. Academic Press.

Pernot, J.-P. (2008). *Incorporating free-form features in aesthetic and engineering product design: State-of-the-art report*. Academic Press.

Cunningham, J., & Dixon, J. (1988). *Designing with features: the origin of features*. In *Proceedings of the 1988 ASME International Computers in Engineering Conference and Exhibition*. ASME.

Wilson, P., & Pratt, P. (1988). A taxonomy of features for solid modeling. Geometric modeling for CAD applications. Elsevier Science Publishers.

Gupta & Gurumoorthy. (2013). *Classification, representation, and automatic extraction of deformation features in sheet metal parts*. Academic Press.

Zhang, Y. (2016). *Feature based building orientation optimization for additive manufacturing*. Academic Press.

Ostergaard & Summers. (2009). *Development of a systematic classification and taxonomy of collaborative design activities*. Academic Press.

Shneiderman & Norman. (1992). *Questionnaire for User Interface Satisfaction (QUIS), Designing the User Interface: Strategies for Effective Human-Computer Interaction*. Academic Press.

Cruz, J. G., Torres, E. M., & Sadek, C. (2015). A Methodology for Modeling and Control of Weld Bead Width in the GMAW Process. *Journal of the Brazilian Society of Mechanical Sciences and Engineering, 37*(5), 1529–1541. doi:10.100740430-014-0299-8

Esakki, B., Rajamani, D., & Arunkumar, P. (2017). Modeling and prediction of optimal process parameters in wear behaviour of selective inhibition sintered high density polyethylene parts. *Progress in Additive Manufacturing, 3*(3), 109–121. doi:10.100740964-017-0033-z

Ngo, T. D., Kashani, A., Imbalzano, G., Nguyen, K. T. Q., & Hui, D. (2018). Additive manufacturing (3D printing): A review of materials, methods, applications and challenges. *Composites. Part B, Engineering, 143*, 172–196. doi:10.1016/j.compositesb.2018.02.012

Richmond, Libonati, & Wettermark. (2017). Detector. *Proceedings of the European Conference on Computer Vision.* . doi:10.6084/m9.figshare.12221543

Sobhi, A., Hamoud, M., & Barakat, A., (2019). Optimal Building Orientation Based on Minimum Volumetric Error using a New Direct Slicing Algorithm. *International Journal of Science and Engineering Investigations, 8*(86).

Wang, C., Tan, X. P., Tor, S. B., & Lim, C. S. (2020). Machine learning in additive manufacturing: State-of-the-art and perspectives. *Additive Manufacturing, 36*, 101538. doi:10.1016/j.addma.2020.101538

Derr, R. L. (1973). *A taxonomy of social purposes of public schools: A handbook*. McKay.

Morris & Stauffer. (1994). *A design taxonomy for eliciting customer requirements*. Academic Press.

Ullman. (1992). *A taxonomy for mechanical design*. Academic Press.

Maidin, Pei, & Campbell. (2019). *A Taxonomy of Design Features for Additive Manufacture*. Academic Press.

Gershenson & Stauffer. (1999). *A taxonomy for design requirements from corporate customers*. Academic Press.

Kemp & Buckner. (1999). *A taxonomy of design guidance for hypermedia design*. Academic Press.

Ding, D., Pan, Z., Cuiuri, D., Li, H., van Duin, S., & Larkin, N. (2016). Bead Modelling and Implementation of Adaptive MAT Path in Wire and Arc Additive Manufacturing. *Robotics and Computer-integrated Manufacturing, 39*, 32–42. doi:10.1016/j.rcim.2015.12.004

Hamoud, M. (2018b). Surface Roughness Justification in Additive Manufacturing. *18th International Conference on Applied Mechanics and Mechanical Engineering-AMME-18*, 185-195. 10.21608/amme.2018.35011

Keshavamurthy, Vijay, & Saravanabavan. (n.d.). Development of Polymer Composites by Additive Manufacturing Process. In *Reference Module in Materials Science and Materials Engineering*. Doi:10.1016/B978-0-12-803581-8.11885-5

Majewski, C. E., Oduye, D., Thomas, H. R., & Hopkinson, N. (2008). Effect of infra-red power level on the sintering behaviour in the high-speed sintering process. *Rapid Prototyping Journal, 14*(3), 155–160. doi:10.1108/13552540810878012

Mandrycky, C., Wang, Z., Kim, K., & Kim, D.-H. (2016). 3D bioprinting for engineering complex tissues. *Biotechnology Advances, 34*(4), 422–434. doi:10.1016/j.biotechadv.2015.12.011 PMID:26724184

Simonyan, K., & Zisserman, A. (2014). *Very deep convolutional networks for large-scale image recognition.* http://purl.fcla.edu/fcla/etd/CFE0006397

Berman, B. (2012). 3-D printing: The new industrial revolution. *Business Horizons, 55*(2), 155–162. doi:10.1016/j.bushor.2011.11.003

Günaydın, K., & Türkmen, H. S. (2018). Common FDM 3D printing defects. *International Congress on 3D Printing (Additive Manufacturing) Technologies and Digital Industry.*

Khoshnevis, B., Asiabanpour, B., Mojdeh, M., & Palmer, K. (2003). SIS - a new SFF method based on powder sintering. *Rapid Prototyping Journal, 9*(1), 30–36. doi:10.1108/13552540310455638

Molero, Fernández, Rodríguez-Alabanda, Guerrero-Vaca, & Romero. (2020). Use of Data Mining Techniques for the Prediction of Surface Roughness of Printed Parts in Polylactic Acid (PLA) by Fused Deposition Modeling (FDM): A Practical Application in Frame Glasses Manufacturing. *Polymers, 2*(12), 840. doi:10.3390/polym12040840

Nele, L., Sarno, E., & Keshari, A. (2013). Modeling of Multiple Characteristics of an Arc Weld Joint. *International Journal of Advanced Manufacturing Technology, 69*(5–8), 1331–1341. doi:10.100700170-013-5077-8

Wang, X., Jiang, M., Zhou, Z., Gou, J., & Hui, D. (2017a). 3D printing of polymer matrix composites: A review and prospective. *Composites. Part B, Engineering, 110*, 442–458. doi:10.1016/j.compositesb.2016.11.034

Ahmad, M. N., Mohd, H. A. R., Maidin, N. A., Osman, M. H., Wahid, M. K., Hussin, M. S. F., & Nur, A. A. A. (2019). Optimization on Surface Roughness of Fused Deposition Modelling (FDM) 3D Printed Parts Using Taguchi Approach. *Symposium on Intelligent Manufacturing and Mechatronics SympoSIMM: Intelligent Manufacturing and Mechatronics*, 230-243.

Lim, C. W. J., Le, K. Q., Lu, Q., & Wong, C. H. (2016). An overview of 3-D printing in manufacturing, aerospace, and automotive industries. *IEEE Potentials, 35*(4), 18–22. doi:10.1109/MPOT.2016.2540098

Rajamani, D., Balasubramanian, E., Arunkumar, P., Silambarasan, M., & Bhuvaneshwaran, G. (2018). Experimental Investigations and Parametric Optimization of Process Parameters on Shrinkage Characteristics of Selective Inhibition Sintered High Density Polyethylene Parts. *Experimental Techniques, 42*(6), 631–644. doi:10.100740799-018-0286-6

Sood, A.K., Ohdar, R., & Mahapatra, S. (2013). Parametric appraisal of mechanical property of fused deposition modelling processed parts. *Materials & Design, 31*(1), 287-95.

Xiong, J., Zhang, G., Hu, J., & Li, Y. (2013). Forecasting Process Parameters for GMAW-Based Rapid Manufacturing Using Closed-Loop Iteration Based on Neural Network. *International Journal of Advanced Manufacturing Technology*, *69*(1–4), 743–751. doi:10.100700170-013-5038-2

Asiabanpour, B., Cano, R., Subbareddy, C., Wasik, F., VanWagner, L., & McCormick, T. (2007). A new heater design by radiation modeling and a new polymer waste-saving mechanism design for the SIS process. *Rapid Prototyping Journal*, *13*(3), 136–147. doi:10.1108/13552540710750889

Cao, Y., Zhu, S., Liang, X., & Wang, W. (2011). Overlapping Model of Beads and Curve Fitting of Bead Section for Rapid Manufacturing by Robotic MAG Welding Process. *Robotics and Computer-integrated Manufacturing*, *27*(3), 641–645. doi:10.1016/j.rcim.2010.11.002

Pérez, M., Medina-Sánchez, G., García-Collado, A., Gupta, M., & Carou, D. (2018). Surface Quality Enhancement of Fused Deposition Modeling (FDM) Printed Samples Based on the Selection of Critical Printing Parameters. *Materials (Basel)*, *11*(8), 1382. doi:10.3390/ma11081382 PMID:30096826

Tambrallimath, Keshavamurthy, Saravanbavan, Pradeepkumar, & Kumar. (2019). Synthesis and Characterization of Graphene Filled PC-ABS Filament for FDM Applications. Advances in Polymer Composites. Mechanics, Characterization and Applications AIP Conf. Proc., 2057.

Williams, C.K.L, Winn, J., & Zisserman, F. (2019). The Pascal Visual Object Classes (VOC) Challenge. International Journal of Computer. *Vision (Basel)*, *88*, 303–338.

Yap, Y., & Yeong, W. (2014). Additive manufacture of fashion and jewellery products: a mini review: This paper provides an insight into the future of 3D printing industries for fashion and jewellery products. *Virtual and Physical Prototyping*, *9*(3), 195–201. doi:10.1080/17452759.2014.938993

Carbonell, M., Fornés, A., Villegas, M., & Lladós, J. (2020). A neural model for text localization, transcription and named entity recognition in full pages. *Pattern Recognition Letters*, *136*, 219–227. doi:10.1016/j.patrec.2020.05.001

Lostado-lorza, R., & Fernández-martínez, R. (2014). *International Joint Conference SOCO'13-CISIS'13-ICEUTE'13*. 10.1007/978-3-319-01854-6

Murphy, S. V., & Atala, A. (2014). 3D bioprinting of tissues and organs. *Nature Biotechnology*, *32*(8), 773–785. doi:10.1038/nbt.2958 PMID:25093879

Raju, M., Gupta, M. K., Bhanot, N., & Sharma, V. S. (2018). A hybrid PSO–BFO evolutionary algorithm for optimization of fused deposition modelling process parameters. *Journal of Intelligent Manufacturing*, *30*(7), 2743–2758. doi:10.100710845-018-1420-0

Tambrallimath, Keshavamurthy, Patil, & Adarsha. (n.d.). Mechanical and Tribological Characteristics of Polymer Composites Developed by Fused Filament Fabrication. *Fused Deposition Modeling Based 3D Printing*, 151-166.

Vahabli, E., & Rahmati, S. (2017). Improvement of FDM parts' surface quality using optimized neural networks medical case studies. *Rapid Prototyping Journal, 23*(4), 825–842. Advance online publication. doi:10.1108/RPJ-06-2015-0075

Boschetto, A., Giordano, V., & Veniali, F. (2011). Modelling micro geometrical profiles in fused deposition process. *International Journal of Advanced Manufacturing Technology, 61*(9-12), 945–956. doi:10.100700170-011-3744-1

Gajdos, Spišák, Kaščák, & Krasinskyi. (2015). Surface Finish Techniques for FDM Parts. *Materials Science Forum, 818*, 45-48.

Jiang, Z., Diggle, B., Tan, M. L., Viktorova, J., Bennett, C. W., & Connal, L. A. (2020). Extrusion 3D printing of polymeric materials with advanced properties. *Advancement of Science, 7*(17), 2001379. doi:10.1002/advs.202001379 PMID:32999820

Joshi, Sheikh, & Prototyping. (2015). *3D printing in aerospace and its long-term sustainability.* Academic Press.

Keshavamurthy, Tambrallimath, Ugrasen, & Girish. (n.d.). Sustainable Product Development by Fused Deposition Modelling Process. *Fused Deposition Modeling Based 3D Printing*, 213-225.

Malviya, R., & Pratihar, D. K. (2011). Tuning of Neural Networks Using Particle Swarm Optimization to Model MIG Welding Process. *Swarm and Evolutionary Computation, 1*(4), 223–235. doi:10.1016/j.swevo.2011.07.001

Akerfeldt, P., Antti, M.-L., & Pederson, R. (2016). Influence of microstructure on mechanical properties of laser metal wire-deposited Ti-6Al-4V. *Materials Science and Engineering A, 674*, 428–437. doi:10.1016/j.msea.2016.07.038

Alabi, M. O., Nixon, K., & Botef, I. (2018). A survey on recent applications of machine learning with big data in additive manufacturing industry. *American Journal of Engineering and Applied Sciences, 11*(3), 1114–1124. doi:10.3844/ajeassp.2018.1114.1124

Alazab, M., Venkatraman, S., Watters, P., & Alazab, M. (2010). *Zero-day malware detection based on supervised learning algorithms of API call signatures.* Academic Press.

Al-Ketan, O., Soliman, A., AlQubaisi, A. M., & Abu Al-Rub, R. K. (2018). Nature-Inspired Lightweight Cellular Co-Continuous Composites with Architected Periodic Gyroidal Structures. *Advanced Engineering Materials, 20*(2), 1700549. Advance online publication. doi:10.1002/adem.201700549

Alzahrani, M., Choi, S. K., & Rosen, D. W. (2015). Design of truss-like cellular structures using relative density mapping method. *Materials & Design, 85*, 349–360. doi:10.1016/j.matdes.2015.06.180

AMFG. (2019). *Application Spotlight: 3D Printing for Footwear.* AMFG. Retrieved from https://amfg.ai/2019/09/18/application-spotlight-3d-printing-for-footwear/

Aviation Pros. (2012). *Turbine Technology: Innovations in Turbine Engines*. Retrieved July 18, 2020, from https://www.aviationpros.com/aircraft/commercial-airline/article/10617063/turbinetechology-innovations-in-turbine-engines

Aviation Pros. (2013). *Stainless Steel: A metal for all seasons*. Retrieved July 18, 2020, from https://www.aviationpros.com/home/article/10387223/stainless-steel-a-metal-forall-seasons

Aviation Safety Network. (2015). *Grumman G-73T Turbo Mallard*. Retrieved July 18, 2020, from https://aviation-safety.net/database/record.php?id=20051219-0

Azzam, R., Taha, T., Huang, S., & Zweiri, Y. (2020). A deep learning framework for robust semantic SLAM. *2020 Advances in Science and Engineering Technology International Conferences, ASET 2020*. 10.1109/ASET48392.2020.9118181

Banga, Gehani, Bhilare, Patel, & Kara. (2018). *3D topology optimization using convolutional neural networks*. arXiv 1808.07440.

Barua, R., Datta, S., Roychowdhury, A., & Datta, P. (2019). Importance of 3D Printing Technology in Medical Fields. In Additive Manufacturing Technologies From an Optimization Perspective (pp. 21-40). IGI Global. doi:10.4018/978-1-5225-9167-2.ch002

Barua, R., Datta, S., Datta, P., & Chowdhury, A. R. (2019). Scaffolds and Tissue Engineering Applications by 3D Bio-Printing Process: A New Approach. In K. Kumar & J. Davim (Eds.), *Design, Development, and Optimization of Bio-Mechatronic Engineering Products* (pp. 78–99). IGI Global. doi:10.4018/978-1-5225-8235-9.ch004

Baturynska, I., Semeniuta, O., & Wang, K. (2018). Application of machine learning methods to improve dimensional accuracy in additive manufacturing. *International Workshop of Advanced Manufacturing and Automation*, 245–252.

Beedham, M. (2019). *This custom 3D-printed bicycle helmet is the future of saving skulls*. TNW. Retrieved from https://thenextweb.com/news/this-custom-3d-printed-bicycle-helmet-is-the-future-of-saving-skulls

Belhocine, A., & Afzal, A. (2020). Computational finite element analysis of brake disc rotors employing different materials. *Australian Journal of Mechanical Engineering*, 1–14. doi:10.1080/14484846.2020.1733175

Bengio, Y., Lamblin, P., Popovici, D., & Larochelle, H. (2006). Greedy layer-wise training of deep networks. *Proceedings of the 19th International Conference on Neural Information Processing Systems*.

Boeing. (2019). *Engine Thrust Hazards in the Airport Environment*. Retrieved July 15, 2020, from https://www.boeing.com/commercial/aeromagazine/aero06/textonly/s02txt.html

Boissonneault, T. (2019). *Northrop Grumman study showcases performance and reliability of 3D printed electronics*. 3D Printing Media Network. Retrieved from https://www.3dprintingmedia.network/northrop-grumman-study-performance-reliability-3d-printed-electronics/

Brambilla, C. R. M., Okafor-Muo, O. L., Hassanin, H., & ElShaer, A. (2021). 3DP Printing of Oral Solid Formulations: A Systematic Review. *Pharmaceutics, 13*(3), 358. https://www.mdpi.com/1999-4923/13/3/358

Bruijne, M. (2016). Machine learning approaches in medical image analysis: From detection to diagnosis. *Medical Image Analysis, 33*, 94–97. doi:10.1016/j.media.2016.06.032 PMID:27481324

Caiazzo, F., & Caggiano, A. (2018). Laser direct metal deposition of 2024 Al alloy: Trace geometry prediction via machine learning. *Materials (Basel), 11*(3), 444. doi:10.3390/ma11030444 PMID:29562682

Chen, D., & Zheng, X. (2018). Multi-material additive manufacturing of metamaterials with giant, tailorable negative Poisson's ratios. *Scientific Reports*, *8*(9139), 9139. doi:10.103841598-018-26980-7 PMID:29904093

Chowdhury, S., Mhapsekar, K., & Anand, S. (2018). Part build orientation optimization and neural network-based geometry compensation for additive manufacturing process. *Journal of Manufacturing Science and Engineering*, *140*(3), 031009. doi:10.1115/1.4038293

Daelemans, W., Hoste, V., Meulder, F. D., & Naudts, B. (2003). Combined optimization of feature selection and algorithm parameters in machine learning of language. *European Conference on Machine Learning*, 84–95. 10.1007/978-3-540-39857-8_10

Datta, S., Barua, R., & Das, J. (2019). *Application of Artificial Intelligence in Modern Healthcare System, Alginates - Recent Uses of This Natural Polymer, Leonel Pereira*. IntechOpen. Available from: https://www.intechopen.com/books/alginates-recent-uses-of-this-natural-polymer/application-of-artificial-intelligence-in-modern-healthcare-system doi:10.5772/intechopen.90454

Davidson, K., & Singamneni, S. (2016). Selective Laser Melting of Duplex Stainless Steel Powders: An Investigation. *Materials and Manufacturing Processes*, *31*(12), 1543–1555. doi:10.1080/10426914.2015.1090605

Debroy, T., Zhang, W., Turner, J., & Babu, S. S. (2017). Building digital twins of 3D printing machines. *Scripta Materialia*, *135*, 119–124. doi:10.1016/j.scriptamat.2016.12.005

Delgoshaei, A., Ariffin, M. K. A. M., Leman, Z., Baharudin, B. T. H. T. B., & Gomes, C. (2016). Review of evolution of cellular manufacturing system's approaches: Material transferring models. *International Journal of Precision Engineering and Manufacturing*, *17*(1), 131–149. doi:10.100712541-016-0017-9

El-Sayed, M. A., Hassanin, H., & Essa, K. (2016). Effect of casting practice on the reliability of Al cast alloys. *International Journal of Cast Metals Research*, *29*(6), 350–354. doi:10.1080/13640461.2016.1145966

Essa, K., Hassanin, H., Attallah, M. M., Adkins, N. J., Musker, A. J., Roberts, G. T., Tenev, N., & Smith, M. (2017). Development and testing of an additively manufactured monolithic catalyst bed for HTP thruster applications. *Applied Catalysis A, General*, *542*, 125–135. doi:10.1016/j.apcata.2017.05.019

Essa, K., Khan, R., Hassanin, H., Attallah, M. M., & Reed, R. (2016). An iterative approach of hot isostatic pressing tooling design for net-shape IN718 superalloy parts. *International Journal of Advanced Manufacturing Technology, 83*(9), 1835–1845. doi:10.100700170-015-7603-3

European Union Aviation Safety Agency. (2019). *Aviation Environmental Impacts*. Retrieved July 6, 2020, from https://www.easa.europa.eu/eaer/climatechange/aviation-environmental-impacts

Faruque, A., Abdullah, M., Chhetri, S. R., Canedo, A., & Wan, J. (2016). Acoustic side-channel attacks on additive manufacturing systems. *Proceedings of the 7th International Conference on Cyber-Physical Systems*, 1–10. 10.1109/ICCPS.2016.7479068

Felzmann, R., Gruber, S., Mitteramskogler, G., Tesavibul, P., Boccaccini, A. R., Liska, R., & Stampfl, J. (2012). Lithography-based additive manufacturing of cellular ceramic structures. *Advanced Engineering Materials, 14*(12), 1052–1058. doi:10.1002/adem.201200010

Gaget, L. (2018). *The world's first 3D printed bike helmet by Kupol!* Sculpteo. Retrieved from https://www.sculpteo.com/blog/2018/10/19/the-world-first-3d-printed-helmet-by-kupol

Galatas, A., Hassanin, H., Zweiri, Y., & Seneviratne, L. (2018). Additive Manufactured Sandwich Composite/ABS Parts for Unmanned Aerial Vehicle Applications. *Polymers, 10*(11), 1262. https://www.mdpi.com/2073-4360/10/11/1262

Ge, W., Lin, F., & Guo, C. (2018). Effect of energy input on microstructure and mechanical properties in EBSM Ti6Al4V. *Materials and Manufacturing Processes, 33*(15), 1708–1713. doi:10.1080/10426914.2015.1048463

Gu, G. X., Chen, C.-T., Richmond, D. J., & Buehler, M. J. (2018). Bioinspired hierarchical composite design using machine learning: Simulation, additive manufacturing, and experiment. *Materials Horizons, 5*(5), 939–945. doi:10.1039/C8MH00653A

Guillame-Gentil, O., Semenov, O., Roca, A. S., Groth, T., Zahn, R., Vörös, J., & Zenobi-Wong, M. (2010). Engineering the extracellular environment: Strategies for building 2D and 3D cellular structures. *Advanced Materials, 22*(48), 5443–5462. doi:10.1002/adma.201001747 PMID:20842659

Gupta, B. B., Tewari, A., Jain, A. K., & Agrawal, D. P. (2017). Fighting against phishing attacks: State of the art and future challenges. *Neural Computing & Applications, 28*(12), 3629–3654. doi:10.100700521-016-2275-y

Hassanin, H., Abena, A., Elsayed, M. A., & Essa, K. (2020). 4D Printing of NiTi Auxetic Structure with Improved Ballistic Performance. *Micromachines, 11*(8), 745. https://www.mdpi.com/2072-666X/11/8/745

Hassanin, H., Zweiri, Y., Finet, L., Essa, K., Qiu, C., & Attallah, M. (2021). Laser Powder Bed Fusion of Ti-6Al-2Sn-4Zr-6Mo Alloy and Properties Prediction Using Deep Learning Approaches. *Materials, 14*(8), 2056. https://www.mdpi.com/1996-1944/14/8/2056

Hassanin, H., Alkendi, Y., Elsayed, M., Essa, K., & Zweiri, Y. (2020). Controlling the Properties of Additively Manufactured Cellular Structures Using Machine Learning Approaches. *Advanced Engineering Materials*, *22*(3), 1901338. doi:10.1002/adem.201901338

Hassanin, H., & Jiang, K. (2010). Optimized process for the fabrication of zirconia micro parts. *Microelectronic Engineering*, *87*(5), 1617–1619. doi:10.1016/j.mee.2009.10.037

Hassanin, H., Modica, F., El-Sayed, M. A., Liu, J., & Essa, K. (2016). Manufacturing of Ti–6Al–4V Micro-Implantable Parts Using Hybrid Selective Laser Melting and Micro-Electrical Discharge Machining. *Advanced Engineering Materials*, *18*(9), 1544–1549. doi:10.1002/adem.201600172

He, K., Yang, Z., Bai, Y., Long, J., & Li, C. (2018). Intelligent fault diagnosis of delta 3D printers using attitude sensors based on support vector machines. *Sensors (Basel)*, *18*(4), 1298. doi:10.339018041298 PMID:29690641

Hinton, G. E., Osindero, S., & Teh, Y.-W. (2006). A Fast Learning Algorithm for Deep Belief Nets. *Neural Computation*, *18*(7), 1527–1554. doi:10.1162/neco.2006.18.7.1527%M16764513

Hinton, G. E., & Salakhutdinov, R. R. (2006). Reducing the Dimensionality of Data with Neural Networks. *Science*, *313*(5786), 504–507. doi:10.1126cience.1127647 PMID:16873662

Hojjati, A., Adhikari, A., Struckmann, K., Chou, E., Nguyen, T. N. T., Madan, K., Winslett, M. S., Gunter, C. A., & King, W. P. (2016). Leave your phone at the door: Side channels that reveal factory floor secrets. *Proceedings of the 2016 ACM SIGSAC Conference on Computer and Communications Security*, 883–894. 10.1145/2976749.2978323

Jafari-Marandi, R., Khanzadeh, M., Tian, W., Smith, B., & Bian, L. (2019). From in-situ monitoring toward high-throughput process control: Cost-driven decision-making framework for laser-based additive manufacturing. *Int. J. Ind. Manuf. Syst. Eng.*, *51*, 29–41. doi:10.1016/j.jmsy.2019.02.005

Jiang, J., Hu, G., Li, X., Xu, X., Zheng, P., & Stringer, J. (2019). Analysis and prediction of printable bridge length in fused deposition modelling based on back propagation neural network. *Virtual and Physical Prototyping*, *14*(3), 253–266. doi:10.1080/17452759.2019.1576010

Jiménez, A., Bidare, P., Hassanin, H., Tarlochan, F., Dimov, S., & Essa, K. (2021). Powder-based laser hybrid additive manufacturing of metals: A review. *International Journal of Advanced Manufacturing Technology*, *114*(1-2), 63–96. Advance online publication. doi:10.100700170-021-06855-4

Kayacan, M. Y., Özsoy, K., Duman, B., Yilmaz, N., & Kayacan, M. C. (2019). A study on elimination of failures resulting from layering and internal stresses in Powder Bed Fusion (PBF) additive manufacturing. *Materials and Manufacturing Processes*, *34*(13), 1467–1475. doi:10.1080/10426914.2019.1655151

Klippstein, H., Hassanin, H., Diaz De Cerio Sanchez, A., Zweiri, Y., & Seneviratne, L. (2018). Additive Manufacturing of Porous Structures for Unmanned Aerial Vehicles Applications. *Advanced Engineering Materials*, *20*(9), 1800290. doi:10.1002/adem.201800290

Körner, C., Attar, E., & Heinl, P. (2011). Mesoscopic simulation of selective beam melting processes. *Journal of Materials Processing Technology, 211*(6), 978–987. doi:10.1016/j.jmatprotec.2010.12.016

Kourou, K., Exarchos, T. P., Exarchos, K. P., Karamouzis, M. V., & Fotiadis, D. I. (2015). Machine learning applications in cancer prognosis and prediction. *Computational and Structural Biotechnology Journal, 13*, 8–17. doi:10.1016/j.csbj.2014.11.005 PMID:25750696

Langford, T., Mohammed, A., Essa, K., Elshaer, A., & Hassanin, H. (2021). 4D Printing of Origami Structures for Minimally Invasive Surgeries Using Functional Scaffold. *Applied Sciences, 11*(1), 332. https://www.mdpi.com/2076-3417/11/1/332

LeCun, Y., Bengio, Y., & Hinton, G. (2015). Deep learning. *Nature, 521*(7553), 436–444. doi:10.1038/nature14539 PMID:26017442

Lee, Park, Cho, Zhang, & Leu. (2001). A neural network approach to the modelling and analysis of stereolithography processes. *Proc. Instn. Mech. Engrs., 215*(Part B), 1719–1733.

Lempitsky, V., & Zisserman, A. (2010). Learning to count objects in images. *Advances in Neural Information Processing Systems*, 1324–1332.

Lewandowski, J. J., & Seifi, M. (2016). Metal additive manufacturing: A review of mechanical properties. *Annual Review of Materials Research, 46*(1), 151–186. doi:10.1146/annurev-matsci-070115-032024

Liang, M., & Hu, X. (2015). Recurrent convolutional neural network for object recognition. *Proceedings of the IEEE Conference on Computer Vision and Pattern Recognition*, 3367–3375. 10.1109/CVPR.2015.7298958

Lim, K. Y. H., Zheng, P., & Chen, C. H. (2019). A state-of-the-art survey of Digital Twin: Techniques, engineering product lifecycle management and business innovation perspectives. *Journal of Intelligent Manufacturing*, 1–25. doi:10.100710845-019-01512-w

Li, S., Da Xu, L., & Zhao, S. (2015). The internet of things: A survey. *Information Systems Frontiers, 17*(2), 243–259. doi:10.100710796-014-9492-7

Li, S., Hassanin, H., Attallah, M. M., Adkins, N. J., & Essa, K. (2016). The development of TiNi based negative Poisson's ratio structure using selective laser melting. *Acta Materialia, 105*, 75–83. doi:10.1016/j.actamat.2015.12.017

Liu, R., Kumar, A., Chen, Z., Agrawal, A., Sundararaghavan, V., & Choudhary, A. (2015). A predictive machine learning approach for microstructure optimization and materials design. *Scientific Reports, 5*(11551), 11551. doi:10.1038rep11551 PMID:26100717

Lu, Y., Liu, C., Kevin, I., Wang, K., Huang, H., & Xu, X. (2020). Digital Twin-driven smart manufacturing: Connotation, reference model, applications and research issues. *Robotics and Computer-integrated Manufacturing, 61*, 101837. doi:10.1016/j.rcim.2019.101837

Materialise. (n.d.a). *The Peugeot Fractal Concept Car: 3D Printing Acoustic Interiors*. Materialise. Retrieved from https://www.materialise.com/en/cases/peugeot-fractal-concept-car-3d-printing-acoustic-interiors

Materialise. (n.d.b). *Philips' Lightbulb Moment: 3D Printing Becomes Essential Production Thinking*. Materialise. Retrieved from https://www.materialise.com/en/cases/philips-lightbulb-moment-3d-printing-becomes-essential-production-thinking

Ma, W., Cheng, F., & Liu, Y. (2018). Deep-learning-enabled on-demand design of chiral metamaterials. *ACS Nano*, *12*(6), 6326–6334. doi:10.1021/acsnano.8b03569 PMID:29856595

McIntosh, T., Jang-Jaccard, J., Watters, P., & Susnjak, T. (2019). The inadequacy of entropy-based ransomware detection. In *International conference on neural information processing*. New York: Springer.

Michaleris, P. (2014). Modeling metal deposition in heat transfer analyses of additive manufacturing processes. *Finite Elements in Analysis and Design*, *86*, 51–60. doi:10.1016/j.finel.2014.04.003

Mohamed, O. A., Masood, S. H., & Bhowmik, J. L. (2017). Influence of processing parameters on creep and recovery behavior of FDM manufactured part using definitive screening design and ANN. *Rapid Prototyping Journal*, *23*(6), 998–1010. doi:10.1108/RPJ-12-2015-0198

Mohammed, A., Elshaer, A., Sareh, P., Elsayed, M., & Hassanin, H. (2020). Additive Manufacturing Technologies for Drug Delivery Applications. *International Journal of Pharmaceutics*, *580*, 119245. doi:10.1016/j.ijpharm.2020.119245 PMID:32201252

Mozaffar, M., Paul, A., Al-Bahrani, R., Wolff, S., Choudhary, A., Agrawal, A., Ehmann, K., & Cao, J. (2018). Data-driven prediction of the high-dimensional thermal history in directed energy deposition processes via recurrent neural networks. *Manufacturing Letters*, *18*, 35–39. doi:10.1016/j.mfglet.2018.10.002

Ning, J., Sievers, D. E., Garmestani, H., & Liang, S. Y. (2019). Analytical modeling of in-process temperature in powder feed metal additive manufacturing considering heat transfer boundary condition. *International Journal of Precision Engineering and Manufacturing-Green Technology*, 1-9.

Ning, F., Cong, W., Qiu, J., Wei, J., & Wang, S. (2015). Additive manufacturing of carbon fiber reinforced thermoplastic composites using fused deposition modeling. *Composites. Part B, Engineering*, *80*, 369–378. doi:10.1016/j.compositesb.2015.06.013

O'Shea, T. J., Corgan, J., & Clancy, T. C. (2016). Convolutional radio modulation recognition networks. *International Conference on Engineering Applications of Neural Networks*, 213–226. 10.1007/978-3-319-44188-7_16

Okafor-Muo, O. L., Hassanin, H., Kayyali, R., & ElShaer, A. (2020). 3D Printing of Solid Oral Dosage Forms: Numerous Challenges With Unique Opportunities. *Journal of Pharmaceutical Sciences*, *109*(12), 3535–3550. doi:10.1016/j.xphs.2020.08.029 PMID:32976900

Okaro, Jayasinghe, Sutcliffe, Black, Paoletti, & Green. (2018). *Automatic Fault Detection for Selective Laser Melting Using Semi-Supervised Machine Learning*. Academic Press.

Olson, G. B. (1997). Computational Design of Hierarchically Structured Materials. *Science*, *277*(5330), 1237–1242. doi:10.1126cience.277.5330.1237

Omar, S., Ngadi, A., & Jebur, H. H. (2013). Machine learning techniques for anomaly detection: An overview. *Int. J. Comput. Appl. Technol. Res.*, *79*(2), 33–41. doi:10.5120/13715-1478

Papastergiou, S., Mouratidis, H., & Kalogeraki, E.-M. (2019). Cyber security incident handling, warning and response system for the european critical information infrastructures (cybersane). In *International Conference on Engineering Applications of Neural Networks* (pp. 476–487). New York: Springer. 10.1007/978-3-030-20257-6_41

Pearson, A. (2020). *How Stratasys 3D printing completely turned around stereo speaker acoustics*. Stratasys. Retrieved from https://www.stratasys.co.in/explore/blog/2016/3d-printed-stereo-speakers

Pham, M. S., Liu, C., Todd, I., & Lertthanasarn, J. (2019). Damage-tolerant architected materials inspired by crystal microstructure. *Nature*, *565*(7739), 305–311. doi:10.103841586-018-0850-3 PMID:30651615

Pilania, G., Wang, C., Jiang, X., Rajasekaran, S., & Ramprasad, R. (2013). Accelerating materials property predictions using machine learning. *Scientific Reports*, *3*(1), 2810. doi:10.1038rep02810 PMID:24077117

Ragan. (2012). *Makerbot Thingverse*. Retrieved from https://www.thingiverse.com/thing:24308

Rahman, N. U., Capuano, L., Cabeza, S., Feinaeugle, M., Garcia-Junceda, A., de Rooij, M. B., Mathews, D. T. A., Walmag, G., Gibson, I., & Römer, G. R. B. E. (2019). Directed energy deposition and characterization of high-carbon high speed steels. *Additive Manufacturing*, *30*(100838). doi:10.1016/j.addma.2019.100838

Rao, P. K. P., Liu, J. P., Roberson, D., Kong, Z. J., & Williams, C. (2015). Online real-time quality monitoring in additive manufacturing processes using heterogeneous sensors. *Journal of Manufacturing Science and Engineering*, *137*(6), 061007. doi:10.1115/1.4029823

Rawat & Shen. (2018). *A novel topology design approach using an integrated deep learning network architecture*. arXiv 1808.02334.

Renken, V., Albinger, S., Goch, G., Neef, A., & Emmelmann, C. (2017). Development of an adaptive, self-learning control concept for an additive manufacturing process. *CIRP Journal of Manufacturing Science and Technology*, *19*, 57–61. doi:10.1016/j.cirpj.2017.05.002

Sabouri, A., Yetisen, A. K., Sadigzade, R., Hassanin, H., Essa, K., & Butt, H. (2017). Three-Dimensional Microstructured Lattices for Oil Sensing. *Energy & Fuels*, *31*(3), 2524–2529. doi:10.1021/acs.energyfuels.6b02850

Scherillo, F., Astarita, A., Carrino, L., Pirozzi, C., Prisco, U., & Squillace, A. (2019). Linear friction welding of Ti-6Al-4V parts produced by electron beam melting. *Materials and Manufacturing Processes*, *34*(2), 201–207. doi:10.1080/10426914.2018.1532086

Schmitt, M., Mehta, R. M., & Kim, I. Y. (2020). Additive manufacturing infill optimization for automotive 3D-printed ABS components. *Rapid Prototyping Journal*, *26*(1), 89–99. doi:10.1108/RPJ-01-2019-0007

Scime, L., & Beuth, J. (2018). Anomaly detection and classification in a laser powder bed additive manufacturing process using a trained computer vision algorithm. *Additive Manufacturing*, *19*, 114–126. doi:10.1016/j.addma.2017.11.009

Shaw, A. (2009). Data breach: From notification to prevention using pci dss. *Colum Soc Probs.*, *43*, 517.

Shi, B., Bai, X., & Yao, C. (2016). An end-to-end trainable neural network for image-based sequence recognition and its application to scene text recognition. *IEEE Transactions on Pattern Analysis and Machine Intelligence*, *39*(11), 2298–2304. doi:10.1109/TPAMI.2016.2646371 PMID:28055850

Silver, D., Schrittwieser, J., Simonyan, K., Antonoglou, I., Huang, A., Guez, A., Hubert, T., Baker, L., Lai, M., Bolton, A., Chen, Y., Lillicrap, T., Hui, F., Sifre, L., Driessche, G. V. D., Graepel, T., & Hassabis, D. (2017). Mastering the game of go without human knowledge. *Nature*, *550*(7676), 354–359. doi:10.1038/nature24270 PMID:29052630

Singh, A., Cooper, D., Blundell, N., Gibbons, G., & Pratihar, D. (2012). Modelling of direct metal laser sintering of EOS DM20 bronze using neural networks and genetic algorithms. *Proceedings of the 37th International MATADOR Conference.*

Sood, A. K., Ohdar, R. K., & Mahapatra, S. S. (2012). Experimental investigation and empirical modelling of FDM process for compressive strength improvement. *Journal of Advanced Research*, *3*(1), 81–90. doi:10.1016/j.jare.2011.05.001

Sosnovik, I., & Oseledets, I. (2019). Neural networks for topology optimization. *Russ. J. Number Anal. M.*, *34*(4), 215–223. doi:10.1515/rnam-2019-0018

Sun, N., Zhang, J., Rimba, P., Gao, S., Zhang, L. Y., & Xiang, Y. (2018). Data-driven cybersecurity incident prediction: A survey. *IEEE Communications Surveys and Tutorials*, *21*(2), 1744–1772. doi:10.1109/COMST.2018.2885561

Sun, X., Bao, J., Li, J., Zhang, Y., Liu, S., & Zhou, B. (2020). A digital twin-driven approach for the assembly-commissioning of high precision products. *Robotics and Computer-integrated Manufacturing*, *61*, 101839. doi:10.1016/j.rcim.2019.101839

Tamburrino, F., Graziosi, S., & Bordegoni, M. (2018). The design process of additively manufactured mesoscale lattice structures: A review. *Journal of Computing and Information Science in Engineering*, *18*(4), 040801. Advance online publication. doi:10.1115/1.4040131

Tanev, H. (2007). Unsupervised learning of social networks from a multiple-source news corpus. *Proceedings of the Workshop Multi-Source Multilingual Information Extraction and Summarization*, 33–40.

Tang, C., Tan, J., & Wong, C. (2018). A numerical investigation on the physical mechanisms of single track defects in selective laser melting. *Int. J. Heat Mass Transf. – Theory Appl.*, *126*, 957–968.

Tan, X. P., Tan, Y. J., Chow, C. S. L., Tor, S. B., & Yeong, W. Y. (2017). Metallic powder-bed based 3D printing of cellular scaffolds for orthopaedic implants: A state-of-the-art review on manufacturing, topological design, mechanical properties and biocompatibility. *Materials Science and Engineering C*, *76*, 1328–1343. doi:10.1016/j.msec.2017.02.094 PMID:28482501

Tong, S., & Koller, D. (2001). Support vector machine active learning with applications to text classification. *Journal of Machine Learning Research*, *2*, 45–66.

Uhlmann, E., Pontes, R. P., Laghmouchi, A., & Bergmann, A. (2017). Intelligent pattern recognition of a SLM machine process and sensor data. *Procedia CIRP*, *62*, 464–469. doi:10.1016/j.procir.2016.06.060

Vyatskikh, A., Delalande, S., Kudo, A., Zhang, X., Portela, C. M., & Greer, J. R. (2018). Additive manufacturing of 3D nano-architected metals. *Nature Communications*, *9*(1), 593. doi:10.103841467-018-03071-9 PMID:29426947

Wang, T., Kwok, T.-H., Zhou, C., & Vader, S. (2018). In-situ droplet inspection and closed-loop control system using machine learning for liquid metal jet printing. *Int. J. Ind. Manuf. Syst. Eng.*, *47*, 83–92. doi:10.1016/j.jmsy.2018.04.003

Ward, L., Agrawal, A., Choudhary, A., & Wolverton, C. (2016). A general-purpose machine learning framework for predicting properties of inorganic materials. *NPJ Comput. Mater.*, *2*, 16028.

Weber, M., Welling, M., & Perona, P. (2000). Unsupervised learning of models for recognition. *European Conference on Computer Vision*, 18–32.

Weichert, D., Link, P., Stoll, A., Rüping, S., Ihlenfeldt, S., & Wrobel, S. (2019). A review of machine learning for the optimization of production processes. *International Journal of Advanced Manufacturing Technology*, *104*(5-8), 1889–1902. Advance online publication. doi:10.100700170-019-03988-5

Wei, H. L., Mazumder, J., & DebRoy, T. (2015). Evolution of solidification texture during additive manufacturing. *Scientific Reports*, *5*(1), 16446. doi:10.1038rep16446 PMID:26553246

Wu, H., Yu, Z., & Wang, Y. (2016). A new approach for online monitoring of additive manufacturing based on acoustic emission. *Proceedings of the ASME 2016 International Manufacturing Science and Engineering Conference*. 10.1115/MSEC2016-8551

Wu, M., Zhou, H., Lin, L. L., Silva, B., Song, Z., Cheung, J., & Moon, Y. (2017). Detecting attacks in Cyber Manufacturing systems: Additive manufacturing example. *MATEC Web of Conferences*, *108*, 06005. doi:10.1051/matecconf/201710806005

Xiong, J., Zhang, G., Hu, J., & Wu, L. (2014). Bead geometry prediction for robotic GMAW based rapid manufacturing through a neural network and a second-order regression analysis. *Journal of Intelligent Manufacturing*, *25*(1), 157–163. doi:10.100710845-012-0682-1

Xu, X., Yang, X., Li, J., Pan, S., Bi, Y., & Gao, Y. (2019). Influence of laser energy distribution on laser surface microstructure processing. *Optik (Stuttgart)*, *199*, 163244. doi:10.1016/j.ijleo.2019.163244

Yan, C., Hao, L., Hussein, A., & Raymont, D. (2012). Evaluations of cellular lattice structures manufactured using selective laser melting. *International Journal of Machine Tools & Manufacture*, *62*, 32–38. doi:10.1016/j.ijmachtools.2012.06.002

Yang, E., Leary, M., Lozanovski, B., Downing, D., Mazur, M., Sarker, A., Khorasani, A. M., Jones, A., Maconachie, T., Bateman, S., Easton, M., Qian, M., Choong, P., & Brandt, M. (2019). Effect of geometry on the mechanical properties of Ti-6Al-4V Gyroid structures fabricated via SLM: A numerical study. *Materials & Design*, *184*, 108165. doi:10.1016/j.matdes.2019.108165

Yang, J., Li, S., Wang, Z., Dong, H., Wang, J., & Tang, S. (2020). Using deep learning to detect defects in manufacturing: A comprehensive survey and current challenges. *Materials (Basel)*, *13*(24), 1–23. doi:10.3390/ma13245755 PMID:33339413

Yang, L., Harrysson, O., Cormier, D., West, H., Gong, H., & Stucker, B. (2015). Additive Manufacturing of Metal Cellular Structures: Design and Fabrication. *JOM*, *67*(3), 608–615. doi:10.100711837-015-1322-y

Yan, W., Qian, Y., Ge, W., Lin, S., Liu, W. K., Lin, F., & Wagner, G. J. (2018). Meso-scale modeling of multiple-layer fabrication process in selective electron beam melting: Interlayer/track voids formation. *Materials & Design*, *141*, 210–219. doi:10.1016/j.matdes.2017.12.031

Yao, X., Moon, S. K., & Bi, G. (2017). A hybrid machine learning approach for additive manufacturing design feature recommendation. *Rapid Prototyping Journal*, *23*(6), 983–997. doi:10.1108/RPJ-03-2016-0041

Ye, D., Hong, G. S., Zhang, Y., Zhu, K., & Fuh, J. Y. H. (2018). Defect detection in selective laser melting technology by acoustic signals with deep belief networks. *International Journal of Advanced Manufacturing Technology*, *96*(5-8), 2791–2801. doi:10.100700170-018-1728-0

Yu, Hur, & Jung. (2018). *Deep learning for topology optimization design.* arXiv 1801.05463.

Yuan, B., Guss, G. M., Wilson, A. C., Hau-Riege, S. P., DePond, P. J., McMains, S., Matthews, M. J., & Giera, B. (2018). Machine-Learning-Based Monitoring of Laser Powder Bed Fusion. *Advanced Materials Technologies*, *3*(12), 1800136. doi:10.1002/admt.201800136

Yuan, S., Shen, F., Bai, J., Chua, C. K., Wei, J., & Zhou, K. (2017). 3D soft auxetic lattice structures fabricated by selective laser sintering: TPU powder evaluation and process optimization. *Materials & Design*, *120*, 317–327. doi:10.1016/j.matdes.2017.01.098

Zhang, X. Z., Leary, M., Tang, H. P., Song, T., & Qian, M. (2018). Selective electron beam manufactured Ti-6Al-4V lattice structures for orthopedic implant applications: Current status and outstanding challenges. *Current Opinion in Solid State and Materials Science*, *22*(3), 75–99. doi:10.1016/j.cossms.2018.05.002

# About the Contributors

**Sachin Salunkhe** is an Associate Professor in the Mechanical Engineering Department at Vel Tech Rangarajan Dr. Sagunthala R&D Institute of Science and Technology Chennai, India. He received his Bachelor's degree in Mechanical Engineering from the Dr. Babasaheb Ambedkar Marathwada University, Aurangabad India. He received his Master of Technology in CAD/CAM and Ph.D from S.V. National Institute of Technology, Surat, India. His main research interests are in the area of press tool design, metal additive manufacturing, AI applications in manufacturing, expert systems for die design, sheet metal forming, wear analysis, fatigue analysis and high strain rate. Dr. Salunkhe has more than 65 research papers have been published in reputed journals, conferences and 05 book chapters (indexed in Web of Science core collection/h-index 6+/ 120+ citations, SCOPUS/h-index 6+/130+ citations, Google Scholar/h-index 9+/145+).He serves as a reviewer for many reputed journals and is a member of the International Association of Engineers (IAENG), and International Association of Computer Science and Information Technology (IACSIT).

**J. Paulo Davim** is a Full Professor at the University of Aveiro, Portugal. He is also distinguished as honorary professor in several universities/colleges/institutes in China, India and Spain. He received his Ph.D. degree in Mechanical Engineering in 1997, M.Sc. degree in Mechanical Engineering (materials and manufacturing processes) in 1991, Mechanical Engineering degree (5 years) in 1986, from the University of Porto (FEUP), the Aggregate title (Full Habilitation) from the University of Coimbra in 2005 and the D.Sc. (Higher Doctorate) from London Metropolitan University in 2013. He is Senior Chartered Engineer by the Portuguese Institution of Engineers with an MBA and Specialist titles in Engineering and Industrial Management as well as in Metrology. He is also Eur Ing by FEANI-Brussels and Fellow (FIET) of IET-London. He has more than 35 years of teaching and research experience in Manufacturing, Materials, Mechanical and Industrial Engineering, with special emphasis in Machining & Tribology. He has also interest in Management, Engineering Education and Higher Education for Sustainability. He has guided

large numbers of postdoc, Ph.D. and master's students as well as has coordinated and participated in several financed research projects. He has received several scientific awards and honors. He has worked as evaluator of projects for ERC-European Research Council and other international research agencies as well as examiner of Ph.D. thesis for many universities in different countries. He is the Editor in Chief of several international journals, Guest Editor of journals, books Editor, book Series Editor and Scientific Advisory for many international journals and conferences. Presently, he is an Editorial Board member of 30 international journals and acts as reviewer for more than 100 prestigious Web of Science journals. In addition, he has also published as editor (and co-editor) more than 200 books and as author (and co-author) more than 15 books, 100 book chapters and 500 articles in journals and conferences (more than 300 articles in journals indexed in Web of Science core collection/h-index 61+/12000+ citations, SCOPUS/h-index 66+/15000+ citations, Google Scholar/h-index 84+/25000+ citations). He has listed in World´s Top 2% Scientists by Stanford University-Elsevier study.

* * *

**Abou Baker Abdel Wahab** is an international consultant expert in mechanical engineering with over 30 years of experience. Specialized in Mechanical Engineering, different fields of technical consultancy. Expert witness in court cases and litigation involving machine deterioration or failure. Professor at faculty of Engineering- Helwan University in Egypt.

**Prafull Agarwal** is B.Tech. in Mechanical Engineering from Manipal University Jaipur. He has authored two papers titled, 'Formulation of Stress Concentration Factor of a Finite Plate with an Elliptical Hole of High Eccentricity Ratio' and 'Evolution of Cryogenics- A Review on Applications of Cryogenics in Medicine'. He has completed a major project with Dr. Ravi Kumar Gupta at Manipal University Jaipur.

**Mohamed Hamoud Ahmed** received his B.Sc. (2002), M.Sc. (2008), and Ph.D. (2015) from Faculty of Engineering, Helwan University, Dept. of Mechanical Engineering. His M.Sc. and Ph.D. in field of Additive Manufacturing (AM). He getting hired as a lecturer in the same Dept. Teaching many courses in the area of CAD/ CAM, AM, Reverse Engineering, Machine Design, Robotics, Non-Conventional Machining, CNC, Engineering Drawing, Metrology, Metal Cutting, Metal Forming, Material handling, theory of machines, CAD software, and CAM software. He published nine papers in AM field and He cooperate as a reviewer for SAGE Journals, Journal of Advances in Mechanical Engineering, in addition, and for 18th International Conference on Applied Mechanics & Mechanical Engineering, 2018,

Military Technical College. He reviewed different articles in the field of AM, Metal Cutting, Artificial Intelligent (AI), Non - Traditional Machining.

**Azza Barakat** received her B.Sc. and M.Sc. from Faculty of Engineering, Helwan University, Dept. of Mechanical Engineering. She was awarded her PhD in Mechanical Engineering from the Imperial College of Science and Technology, University of London, from which she also obtained a DIC Diploma. The internationalization of her experience is also reflected on her postgraduate students as evident in supervising more than thirty M.Sc. and PhD theses. Her publications enriched research in the field of Production Engineering. With more than fifty publications in the field, she succeeded in guiding and orienting scholars. She participated at various local and global conferences where her expertise in areas such as: CAD/CAM, Sheet Metal Cutting Die Design, Algorithm and Computer-aided Design. She taught numerous undergraduate and postgraduate courses like Computer-aided Manufacturing, Material Handling Equipment, Maintenance Management and Technology, Environmental Studies, Engineering Economy and Reverse Engineering. She also has a very good knowledge of Harvard Graphics, Primavera, Maint Scape and the various packages for CAM software.

**Balasubramanian Esakki** has completed his PhD from Concordia University Canada. Currently he is working as a Professor in the Department of Mechanical Engineering at Vel Tech Rangarajan Dr. Sagunthala R&D Institute of Science and Technology, India. His research interests include micro aerial vehicles, additive manufacturing and automation.

**Ravi Kumar Gupta**, Associate Professor in the Department of Mechanical Engineering at Manipal University Jaipur and at Manipal Institute of Technology Bengaluru, Manipal Academy of Higher Education, Manipal, India; received Ph.D. from IISc Bangalore and M.Tech. from MNNIT Allahabad. He also pursued a postdoc program at Ecole Central de Nantes, France, being a partner of ProRegio project. He has 16 years Industry-Academia blended experience and has published more than 50 research papers. Research Interest includes Product Development; 3D Digital Modelling; CAD/CAM; Product Informatics; Reverse Engineering; Additive Manufacturing; Augmented and Virtual Reality.

**Siva Kumar M.** is currently working as a Professor in the Department of Mechanical Engineering at Vel Tech Rangarajan Dr. Sagunthala R&D Institute of Science and Technology, India. He obtained his PhD degree from Anna University. He has more than 50 articles in referred journals on his credit. His research interests including optimization, meta-heuristic algorithms, and manufacturing systems.

**V. Mohanavel** is working as Associate Professor in Bharath Institute of Higher Education and Research, Chennai, Tamil Nadu, India. He graduated his B.E degree in Mechanical Engineering from Anna University Affiliated College in Vellore, India. He received his M.Tech degree in Design Engineering from Dr.M.G.R Educational and Research Institute University, Chennai, Tamil Nadu, India. He obtained his Ph.D degree in Mechanical Engineering Department from St.Peter's Higher Education and Research, Chennai. He is doing his research work in the area of Composite Materials, Powder Metallurgy Techniques, Stir Casting Process, Process optimization, Polymers, Welding Technology and Material Synthesis & Characterization. He also is engaging with Bharath Institute of Higher Education and Research and Anna University, Chennai as a recognized supervisor in the faculty of Mechanical Engineering for guiding Ph.D. (By Research). He has published more than 150 papers in various refereed international journals. He has presented more than 80 papers in International and national conferences. He is a life member of the Indian Society for Technical Education (ISTE) and Institution of Engineers India (IEI).

**Vishal Naranje** is an Associate professor in the Department of Mechanical Engineering at Amity University Dubai. He had 20 years of teaching and research experience in Indian and foreign universities. His research interest concerns mainly in Design of Expert Systems, Digital Manufacturing, Decision making tools, Micro-machining. He has published more than 50 research articles, books chapters in SCI, Scopus indexed journals and conferences etc. He has received research grants from various funding agencies in India. He has completed few industrial consultancy projects in India and UAE. He is reviewer for the journals published by Springer, Elsevier, Taylor and Francis, etc.

**D. Rajamani** is currently working as an Associate Professor in the Department of Mechanical Engineering at Vel Tech Rangarajan Dr. Sagunthala R&D Institute of Science and Technology, India. He has completed his Undergraduate and Postgraduate degrees from Anna University, India o 2012 & 2014, respectively. He carried out his Doctoral research work in the field of Development of a cost-effective additive manufacturing system for near net shape polymer parts, which was supported by Defence Research and Development Organization, Government of India. He has published more than 30 articles in reputed Journals.

**C. Rajkumar** received his Bachelor of Engineering degree in 1999 with First Class in Production Engineering from St. Peter's Engineering College affiliated to University of Madras, Chennai, Tamil Nadu, India. He received his Master of Engineering (Manufacturing) degree in 2006 from MIT Campus, Anna University, Chennai, Tamil Nadu, India. He started his career as a Lecturer in Vel Tech Poly-

technic College, Chennai, Tamil Nadu, India, in the year 2001 and now working as a Associate Professor in the Department of Mechanical Engineering in Vel Tech Rangarajan Dr. Sagunthala R&D Institute of Science and Technology. He has more than twenty years of teaching experience. He taught subjects such as Manufacturing Technology, Machining and Machine Tools Technology, Engineering Drawing, Engineering Materials and Metallurgy, Project Management and Finance as well as Fuels and Combustion courses for the UG students. He has published sixteen papers in International Journals.

**Atiq Ur Rehman** is a keen student of life and a firm believer in growth through impact. He is a true startup growth hacker with an eye for innovation and a background in Digital Twin prototyping, marketing strategy & execution, brand strategy, UI UX design/research and business development / consulting. There could be millions of ways to solve for an issue, but they all start from the same fundamental understanding of how input processes and outputs work. Between that thought and reality is where his spark for innovation lies. Atiq has experience from many industries such education, technology, engineering, and the arts; almost exclusively from startups because he thrives on their fast-paced, challenging, high-impact environment. Throughout all this, he has been driven by a passion to leverage his skills and knowledge to benefit others, for example by co-creating an alternative learning method for youth that makes them (actually) future-ready.

**Kurian Rishi**, B.Tech. in Mechanical Engineering from Manipal University Jaipur, India. He has undertaken a GIAN course on Micro and Nano Manufacturing at NIT Calicut and completed a major and mini-project with Dr. Ravi Kumar Gupta at Manipal University Jaipur.

**Jayavelu S.** obtained his Bachelor of Engineering with First class in Mechanical Engineering from Thanthai Periyar Government Institute of Technology, Vellore in the year 1998, Master of Engineering (Computer Aided Design) first class with distinction from Satyabama University, Chennai (India) in the year 2010 and Doctor of Philosophy from Vel Tech Rangarajan Dr. Sagunthala R&D Institute of Science and Technology, Chennai in the year 2020 in the area of Aluminium Hybrid metal matrix composite materials. He has done his Schooling from Vellayan Chettiyar Higher Secondary School, Thiruvottiyur, Chennai and St. Joseph's Higher Secondary School, Ennore. He got centum in mathematics in tenth standard a state level examination of the Tamil Nadu. He joined as Lecturer in Vel Tech Rangarajan Dr. Sagunthala R&D Institute of Science and Technology, Chennai and at present holding a post of Associate Professor in the Mechanical Department. He has 12 years of experience in teaching filed and handled various courses for undergradu-

ate and post graduate students. Guided number of projects for the undergraduate students and presented papers in national and international conferences in area of composite materials.

# Index

## O



## P



## R



## S



## T



## W